ANIMAL AND PLANT
Anatomy

VOLUME CONSULTANTS

• Deborah Bodolus, *East Stroudsburg University, PA* • Amy-Jane Beer, *Natural history writer and consultant*
• Kieran Pitts, *Bristol University, England* • Phil Whitfield, *King's College, London*

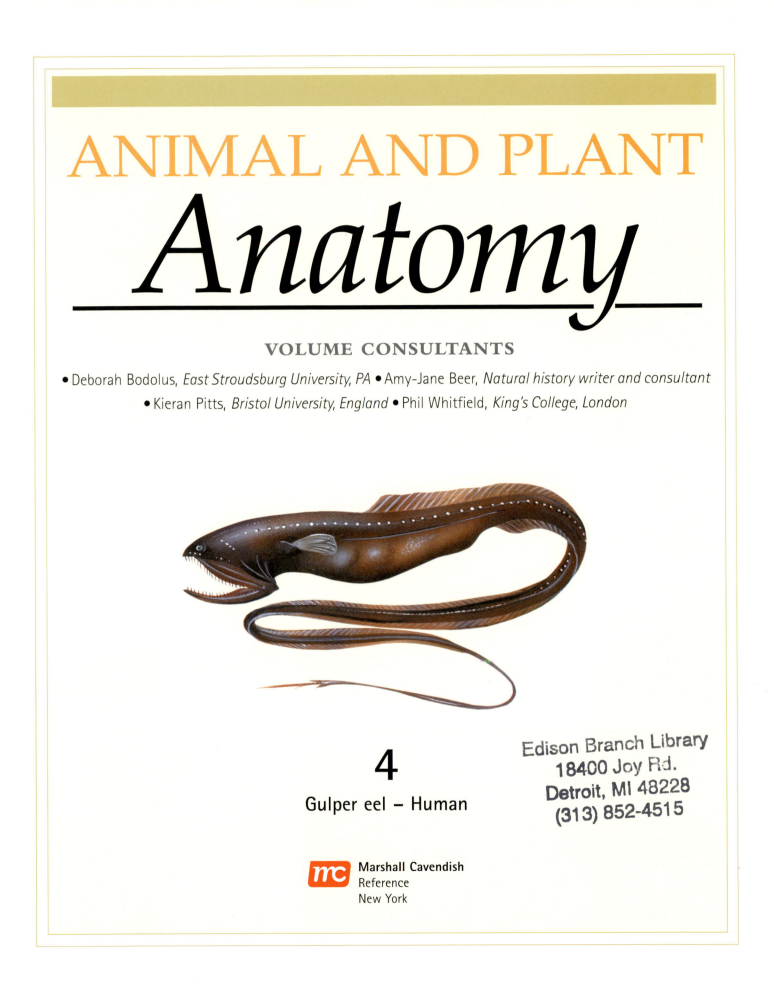

4

Gulper eel – Human

mc **Marshall Cavendish**
Reference
New York

CONTRIBUTORS

Roger Avery; Richard Beatty; Amy-Jane Beer; Erica Bower; Trevor Day; Erin Dolan; Bridget Giles; Natalie Goldstein; Tim Harris; Christer Hogstrand; Rob Houston; John Jackson; Tom Jackson; James Martin; Chris Mattison; Katie Parsons; Ray Perrins; Kieran Pitts; Adrian Seymour; Steven Swaby; John Woodward.

CONSULTANTS

Barbara Abraham, Hampton University, VA; Glen Alm, University of Guelph, Ontario, Canada; Roger Avery, Bristol University, England; Amy-Jane Beer, University of London, England; Deborah Bodolus, East Stroudsburg University, PA; Allan Bornstein, Southeast Missouri State University, MO; Erica Bower, University of London, England; John Cline, University of Guelph, Ontario, Canada; Trevor Day, University of Bath, England; John Friel, Cornell University, NY; Valerius Geist, University of Calgary, Alberta, Canada; John Gittleman, University of Virginia, VA; Tom Jenner, Academia Británica Cuscatleca, El Salvador; Bill Kleindl, University of Washington, Seattle, WA; Thomas Kunz, Boston University, MA; Alan Leonard, Florida Institute of Technology, FL; Sally-Anne Mahoney, Bristol University, England; Chris Mattison; Andrew Methven, Eastern Illinois University, IL; Graham Mitchell, King's College, London, England; Richard Mooi, California Academy of Sciences, San Francisco, CA; Ray Perrins, Bristol University, England; Kieran Pitts, Bristol University, England; Adrian Seymour, Bristol University, England; David Spooner, University of Wisconsin, WI; John Stewart, Natural History Museum, London, England; Erik Terdal, Northeastern State University, Broken Arrow, OK; Phil Whitfield, King's College, University of London, England.

Marshall Cavendish

99 White Plains Road
Tarrytown, NY 10591–9001

www.marshallcavendish.us

Library of Congress Cataloging-in-Publication Data
Animal and plant anatomy.
    p. cm.
  ISBN-13: 978-0-7614-7662-7 (set: alk. paper)
  ISBN-10: 0-7614-7662-8 (set: alk. paper)
  ISBN-13: 978-0-7614-7667-2 (vol. 4)
  ISBN-10: 0-7614-7667-9 (vol. 4)
  1. Anatomy. 2. Plant anatomy. I. Marshall Cavendish Corporation. II.
Title.

  QL805.A55 2006
  571.3--dc22

                                        2005053193

Printed in China
09 08 07 06 1 2 3 4 5

MARSHALL CAVENDISH
Editor: Joyce Tavolacci
Editorial Director: Paul Bernabeo
Production Manager: Mike Esposito

THE BROWN REFERENCE GROUP PLC
Project Editor: Tim Harris
Deputy Editor: Paul Thompson
Subeditors: Jolyon Goddard, Amy-Jane Beer, Susan Watts
Designers: Bob Burroughs, Stefan Morris
Picture Researchers: Susy Forbes, Laila Torsun
Indexer: Kay Ollerenshaw
Illustrators: The Art Agency, Mick Loates, Michael Woods
Managing Editor: Bridget Giles

Contents

Gulper eel

ORDER: Saccopharyngiformes FAMILY: Saccopharyngidae
GENUS: *Saccopharynx*

The common Atlantic gulper eel or swallower, *Saccopharynx ampullaceus*, lives in the Atlantic Ocean at depths of 3,300 to 10,000 feet (about 1,000–3,000 m). With its bioluminescent (light-producing) tail probably acting as a lure to attract prey, the eel feeds on fish and crustaceans. It is capable of consuming very large prey, since it has a cavernous mouth and a distensible (stretchable) stomach.

Anatomy and taxonomy

Scientists categorize all organisms into taxonomic groups based partly on anatomical features. The Atlantic gulper eel is one of at least nine species of swallowers in the family Saccopharyngidae. Together with the pelican eels, single-jaw eels, and bobtail snipe eels, they form a group of eel-like jawed fish in the order Saccopharyngiformes.

- **Animals** Gulper eels, like other animals, are multicellular and gain their food supplies by consuming other organisms. Animals differ from other multicellular life forms in their ability to move from one place to another (in most cases, using muscles). They generally react rapidly to touch, light, and other stimuli.

- **Chordates** At some time in its life cycle a chordate has a stiff, dorsal (back) supporting rod called the notochord that runs all or most of the length of the body.

- **Vertebrates** In all living vertebrates except hagfish, the notochord develops into a backbone (spine, or vertebral column) made up of units called vertebrae. The vertebrate muscular system that moves the body consists primarily of muscles in a mirror-image arrangement on either side of the backbone or the notochord. This arrangement is called bilateral symmetry.

- **Gnathostomes** Gnathostomes are jawed vertebrate fish, unlike hagfish and lampreys (agnathans), which are fish that lack true jaws. Gnathostomes have gills that open to the outside through slits, and pectoral and pelvic fins.

▶ *This family tree shows the major groups of living deep-sea eels, including the gulpers and swallowers. The pelican eel is sometimes called a gulper eel, but the common Atlantic gulper eel is in the family Saccopharyngidae.*

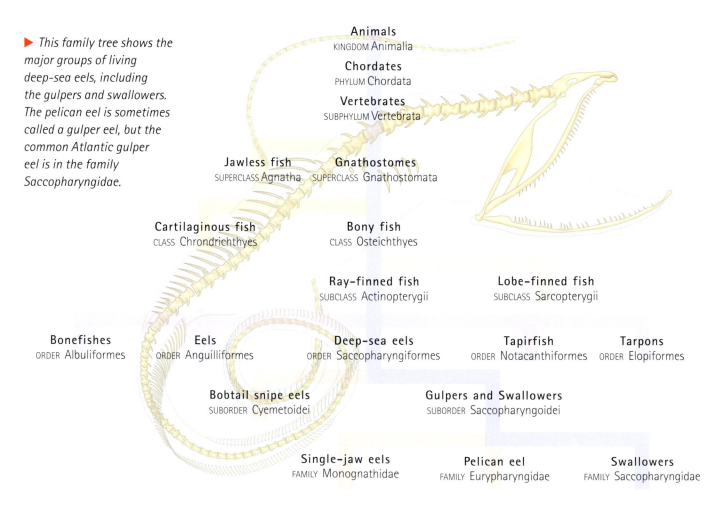

Animals
KINGDOM Animalia

Chordates
PHYLUM Chordata

Vertebrates
SUBPHYLUM Vertebrata

Jawless fish
SUPERCLASS Agnatha

Gnathostomes
SUPERCLASS Gnathostomata

Cartilaginous fish
CLASS Chrondrichthyes

Bony fish
CLASS Osteichthyes

Ray-finned fish
SUBCLASS Actinopterygii

Lobe-finned fish
SUBCLASS Sarcopterygii

Bonefishes
ORDER Albuliformes

Eels
ORDER Anguilliformes

Deep-sea eels
ORDER Saccopharyngiformes

Tapirfish
ORDER Notacanthiformes

Tarpons
ORDER Elopiformes

Bobtail snipe eels
SUBORDER Cyemetoidei

Gulpers and Swallowers
SUBORDER Saccopharyngoidei

Single-jaw eels
FAMILY Monognathidae

Pelican eel
FAMILY Eurypharyngidae

Swallowers
FAMILY Saccopharyngidae

● **Bony fish** Gulper eels belong to the class Osteichthyes (bony fish), the major group that includes more than 95 percent of all fish. Bony fish, as their name implies, have a skeleton of bone, as opposed to members of the class Chondrichthyes (cartilaginous fish such as sharks, skates, and rays), which have a skeleton made of cartilage.

● **Ray-finned fish** Almost all bony fish, including gulper eels, belong to the subclass Actinopterygii (ray-finned fish). The major feature that distinguishes them from the eight species of the subclass Sarcopterygii (lobe-finned fish) is the presence of bony rays that support thin, flexible fins.

● **Deep-sea eels** During the long process of evolution, members of the order Saccopharyngiformes have lost or reduced many features found in other ray-finned fish. Food is generally scarce at great depths, and deep-sea eels cannot waste any opportunities to catch prey. Some have evolved large jaws and a capacious stomach, so almost any prey they encounter can be eaten. Most saccopharyngiform fish keep energy consumption to a minimum; their muscles and skeletons are small and simple compared with those of shallow-water fish. Saccopharyngiform fish are eel-like in body form, with long dorsal and anal fins.

● **Bobtail snipe eels** The bobtail snipe eels or arrow eels are the only members of the suborder Cyematoidei. Bobtail snipe eels grow to only 6 inches (15 cm) long. Their dorsal and anal fins extend back as far as the tail, and are cut off like the feathers of a dart or arrow.

● **Single-jaw eels** Like swallower eels, single-jaw eels have a mouth with a very large gape and an expandable stomach. Unlike all other deep-sea eels they lack an upper jaw and pectoral fins. Uniquely among fish, the single-jaw eels have a hollow fang in the roof of the mouth that points forward. The fang is supplied with venom glands in a manner similar to the venom sacs of some snakes.

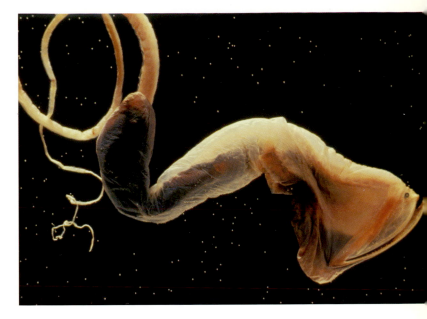

▲ *The common Atlantic gulper eel is one of nine species in the swallower eel family. It lives at great ocean depths; its most obvious features are its huge jaws and extremely long tail.*

● **Pelican eel** The single species *Eurypharynx pelecanoides* has a huge mouth—even larger than that of swallower eels (below)—with jaws armed with numerous tiny teeth. The six pairs of gill openings are covered by a flap of skin, the operculum, that is not supported by bone. The pelican eel's pectoral fins are minute. As in swallower eels, the tiny eyes are set near the tip of the snout. The pelican eel's elongated body, which grows to about 24 inches (60 cm) in length, contains 100 to 125 vertebrae. The long tail has a bio-luminescent organ near the tip.

● **Swallowers** The nine species of swallower eels, including the Atlantic gulper eel, have large mouths armed with curved teeth. The pectoral fins are of moderate size. The body is enormously elongated, with a backbone containing 150 to 300 vertebrae, and a very long, slender tail bearing a bioluminescent organ.

FEATURED SYSTEMS

EXTERNAL ANATOMY Features include a mouth with an enormous gape, a saclike abdomen, and a long tail with a bioluminescent organ near its tip. *See pages 438–440.*

SKELETAL AND MUSCULAR SYSTEMS These systems are relatively small and simple in structure. *See page 441.*

NERVOUS SYSTEM The most developed features are those concerned with the detection of vibrations (the lateral line system) and smell (the olfactory system). *See page 442.*

CIRCULATORY AND RESPIRATORY SYSTEMS The heart and gills are comparatively small, as befits a fish living a sluggish life in an oxygen-rich environment. Red blood cells make up only 10 percent of the blood. *See page 443.*

DIGESTIVE AND EXCRETORY SYSTEMS The mouth and stomach are capable of engulfing prey larger in mass and volume than the gulper eel itself. *See page 444.*

REPRODUCTIVE SYSTEM Breeding males and females devote their energy to reproduction. They forgo food and die after mating. The female releases thousands of eggs that are fertilized externally and float to near-surface waters to hatch as leptocephalus ("leaflike") larvae. *See page 445.*

External anatomy

COMPARE a gulper eel's very small pectoral fins with those of a *SAILFISH*.

COMPARE the cavernous jaws of a gulper eel with the relatively small jaws of a *TROUT*.

Gulper eels in the genus *Saccopharynx* are classic deep-sea fish. They have a cavernous mouth, an elastic stomach, toothed jaws, and a luminous organ on the tail. The common Atlantic gulper eel, like other species in this genus, is uniformly black. At the great depths at which it lives (below 3,300 ft., or 1,000 m), there is no natural light other than bioluminescence, the light produced by fish and other creatures. The gulper eel's dark skin is effective camouflage in a dark environment.

The gulper eel's head is dominated by an enormous mouth armed with backward-pointing curved teeth for grasping prey. *Saccopharynx* gulper eels feed predominantly on other fish and crustaceans, and their jaws can open sufficiently wide to consume prey

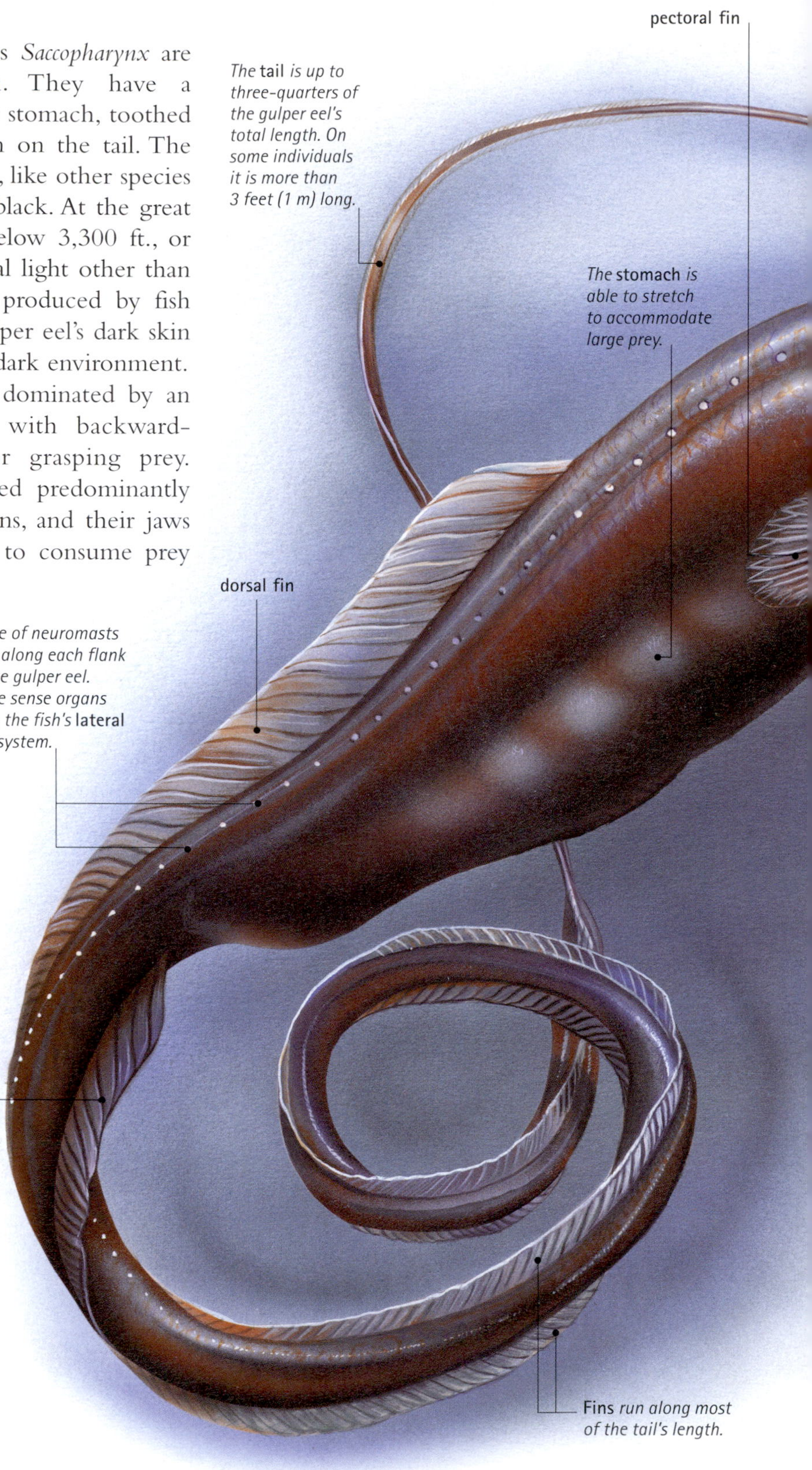

▶ **Common Atlantic gulper eel**
This individual has an impressive set of teeth. Gulpers lose their teeth when they are ready to breed. In both sexes, the tail makes up more than half the total length of the fish.

COMPARATIVE ANATOMY

The pelican eel

The pelican eel *Eurypharynx pelecanoides* is sometimes called a gulper eel. It is smaller than the *Saccopharynx* gulper eels, but has a mouth that is even larger in proportion to its body size. The fish gains its name from the elastic pouch hanging from its lower jaw. The pelican eel has smaller teeth than *Saccopharynx* gulpers and, as is known from the stomach contents of caught specimens, the pelican eel eats small shrimp. The pelican eel probably swims slowly through the water with its mouth open, engulfing small prey.

pectoral fin

The tail is up to three-quarters of the gulper eel's total length. On some individuals it is more than 3 feet (1 m) long.

The stomach is able to stretch to accommodate large prey.

dorsal fin

A line of neuromasts runs along each flank of the gulper eel. These sense organs form the fish's lateral line system.

ventral fin

Fins run along most of the tail's length.

tail
filaments

The small
**bioluminescent
organ** of a gulper
eel flashes light.
This probably
attracts prey.

Each of the two
eyes is tiny in
relation to the total
length of the fish
but is relatively
large in relation
to the skull.

Two **nostrils**
connect to the
olfactory, or
smell, organ.

The **teeth** are long
and sharp. They
point forward
when the mouth
is closed but
backward when
the mouth opens.

The **teeth** of the
lower jaw are
slightly shorter
than those of
the upper jaw.

The **mouth** can
open very wide.
The backward-
pointing teeth
keep prey from
escaping.

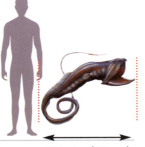

80 inches (200 cm)

IN FOCUS

Producing light

Many deep-water fish and invertebrates
produce their own light. It is called
bioluminescence. Fish produce light in two
ways. Some species, such as lanternfish (order
Myctophiformes), have modified mucus
glands that contain chemicals called luciferins.
The luciferins undergo chemical changes to
release light energy. Such fish manufacture
their own luciferins and luciferase (an enzyme
vital to the release of light), and light emission
is controlled by their nervous system. So,
they are able to flash light on and off in
complicated ways. Other fish, including
gulper eels, have organs that contain
symbiotic bacteria. It is the bacteria that
contain luciferins and emit light. In some
species, the bacteria can be stimulated by the
fish's nervous system to flash, but the control
is less precise than in the case of those fish
that produce their own light.

that are larger in mass and volume than the
gulper eel itself. The gulper eel's throat and
stomach are elastic and able to stretch to
accommodate such large meals. As the food
becomes digested inside the gut, the fish's belly
gradually reduces. A single large meal can keep
the gulper eel satisfied for many weeks.

IN FOCUS

The world of the gulper eel

Gulper eels live in the largest near-uniform environment on Earth.
The midwater zone of the oceans, between about 3,300 feet and 16,500
feet (1,000–5,000 m), is called the bathypelagic zone (from the Greek
bathys for "depth" and *pelagios* for "of the sea"). This zone makes up
about half the ocean's water and contains more than one-third of the
living space on the planet. The bathypelagic zone has conditions that
seem very alien to humans. The water pressure there is 100 to 500 times
greater than atmospheric pressure at the surface. No sunlight reaches
the bathypelagic zone, and the only light there is produced by the
animals themselves. The temperature is a near-constant 34°F to 39°F
(1°C–4°C), and the concentrations of chemicals, including salt and
oxygen, remain more or less constant. Very little food reaches the
bathypelagic zone from the productive waters above, and so animals
in this zone are relatively scarce.

The role of bioluminescence

Bioluminescence probably serves several functions. In some species, such as deep-sea angler fish (order Lophiiformes) and probably gulper eels, a light organ is used as a lure to attract prey close to the mouth. In other fish, such as the lanternfish (order Myctophidae), the light patterns identify individuals of the same species in the darkness. Hatchetfish live in the twilight zone (at depths of about 660 to 3,300 ft., or 200 to 1,000 m). Light organs on their underside break up the silhouette of the fish against the dim light streaming through the water from above. Thus, the lights serve to camouflage the hatchetfish from predators hunting below.

BIOLUMINESCENT TAIL
A gulper eel's light organ contains bacteria that have luciferins, which emit light. The bioluminescent section is less than 0.5 inch (1.2 cm) long.

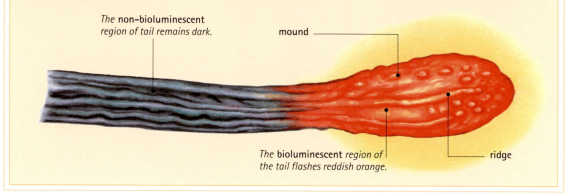

The **non-bioluminescent** region of tail remains dark.

mound

The **bioluminescent** *region of the tail flashes reddish orange.*

ridge

The gills are positioned farther back on the body of the gulper eel than they are on other ray-finned fish: they are behind the fish's enormous mouth. The plumelike gills are covered by a boneless flap of skin, the operculum.

The dorsal (back) and ventral (underside) fins of a gulper eel are elongated, but the muscles of the body and tail are so small (they are reduced relative to other ray-finned fish) that the fish swims only sluggishly. Gulper eels probably remain hanging in the water almost motionless for much of the time, waiting for prey to approach. The forward-seeing eyes on the fish's snout are relatively small, but they are sufficient to pick out flashes of bioluminescent light in the gloom.

Light in the gloom
One of the gulper eel's most curious features is its tapering, whiplike tail, which extends behind the anus. The tail makes up about three-quarters of the total length of the body and has a flashing luminescent organ near its tip. In freshly caught specimens, the organ (an elongated arrangement of tentacles, mounds, thin projections called papillae, and grooves) emits a continuous pink glow, punctuated by flashes of red. Its function is unclear. Some scientists believe that the organ might act as a decoy to misdirect predators. However, most ichthyologists (scientists who study fish) think it is a lure to tempt prey toward the jaws. For this ploy to work, the fish would need to dangle its tail in front of its mouth.

Diversity in the deep ocean

What the bathypelagic zone lacks in fish abundance it makes up for in diversity. Apart from gulper eels, more than 100 species of anglerfish, plus dozens of bristlemouths, blackdragons, and viperfish, are known inhabitants of this alien world. Many others probably remain to be discovered by scientists. Like most gulper eels, many of these deep-ocean fish have a gaping mouth and a distensible stomach. These similarities are examples of convergent evolution. Convergent evolution is the process whereby different, and not closely related, animals evolve superficially similar structures in response to the demands of living a comparable lifestyle in the same environment.

Skeletal and muscular systems

Gulper eels do not have a very streamlined shape. Coupled with their feebly developed body muscles, this lack of streamlining suggests they are poor swimmers. The skeleton of a gulper eel is neither as strong nor as complex as those of shallow-water, ray-finned fish. The skeleton is only weakly impregnated with calcium salts, so gulper eel bones are relatively fragile. The high-pressure environment of the deep ocean provides the fish with all the support its body requires. Gulper eels do not swim rapidly to catch their prey and so do not need large, strong bones on which to anchor large muscles. These fish also lack scales; their body does not require the kind of skin protection or streamlining that is required by fast swimmers.

The skull of gulper eels is extremely small, typically less than 2 percent of the length of the animal's body. However, the jaws are very long. Those of the common Atlantic gulper eel and other species in the genus *Saccopharynx* are about between one-quarter and one-third the length of the trunk. A structure called the suspensorium, which is hinged against the cranium and from which the lower jaw hangs, is also very elongated.

The jaws are hinged in such a way that they have an enormous gape. The jawbones are attached to the cranium by elastic ligaments, which permit movement of the bones through very large angles. Unusually for most fish, when a gulper eel's mouth is closed it slopes downward from front to back at an angle of about 45 degrees, and the teeth in the top jaw are turned upward. When the lower jaw swings downward to open the mouth, the teeth in the upper jaw swing and rotate into a biting position. The jaw muscles are the only well-developed muscles in a gulper eel's body.

CONNECTIONS

COMPARE a gulper eel's jaw mechanism with that of a *TROUT* and with the jawless head of a *HAGFISH*.

COMPARE the weakly developed muscles of a gulper eel with those of a fast-swimming fish, such as a *HAMMERHEAD SHARK* or a *SAILFISH*.

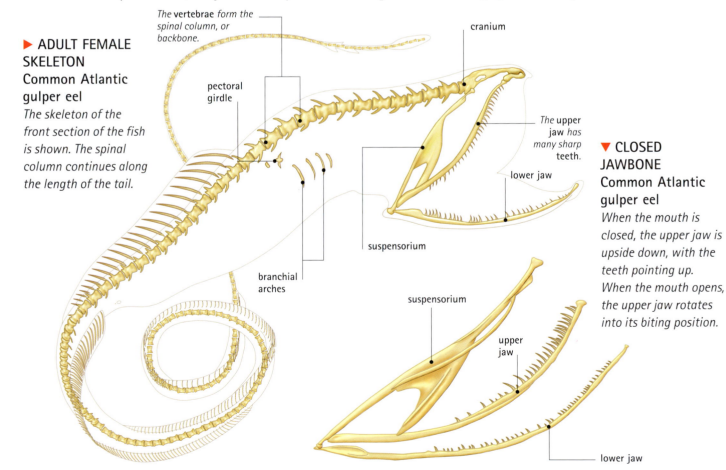

▶ **ADULT FEMALE SKELETON**
Common Atlantic gulper eel
The skeleton of the front section of the fish is shown. The spinal column continues along the length of the tail.

*The **vertebrae** form the spinal column, or backbone.*

cranium

pectoral girdle

*The **upper jaw** has many sharp teeth.*

lower jaw

suspensorium

branchial arches

▼ **CLOSED JAWBONE**
Common Atlantic gulper eel
When the mouth is closed, the upper jaw is upside down, with the teeth pointing up. When the mouth opens, the upper jaw rotates into its biting position.

suspensorium

upper jaw

lower jaw

Nervous system

CONNECTIONS

COMPARE the lateral line system of a gulper eel, particularly the way it divides on the head, with that of a *HAMMERHEAD SHARK* and a *TROUT*.

The nervous system of gulper eels is based on the same plan as that of other vertebrates. The central nervous system (CNS) consists of the brain and the spinal cord. The CNS is connected to the peripheral nervous system (PNS), which consists of nerves that lead to and from sensory organs and to responsive structures such as muscles.

The brain is broadly divided into three regions, as it is in other vertebrates: the fore-, hind-, and midbrain. A gulper eel's brain is smaller than that of most other fish, and most parts of the nervous system are relatively simple. However, those regions of the CNS, PNS, and sensory organs that are associated with the lateral line (vibration-detecting) system and olfactory (odor-detecting) organs are more complex.

▼ CRANIAL NERVES
Pelican eel

This is a view from below of the eel's head. The anterior of the brain is at the top of the diagram. The cranial nerve plan of gulper eels is similar.

Gulper eel senses

The most important sense in gulper eels is probably the ability to detect vibrations. Vibrations are caused by disturbances in the water. The main means of sensing the disturbances is through the lateral line system. In most fish that live in shallow water this system consists of cells called neuromasts, from

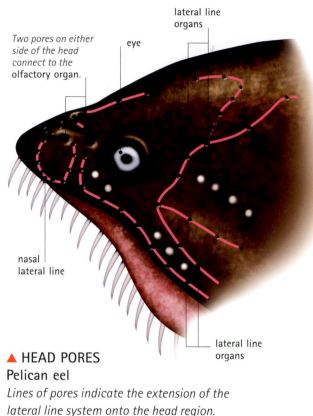

▲ HEAD PORES
Pelican eel

Lines of pores indicate the extension of the lateral line system onto the head region. Such an arrangement is similar for gulper eels.

which gel-filled canals extend to the surface of the fish. The canals reach the surface in a lateral line that runs along the fish's flanks. The arrangement in gulper eels is different, since the neuromasts protrude a little way from the lateral line and dangle in the water. Organized in this way, the neuromasts are ultrasensitive to vibrations in the water.

The eyes of *Saccopharynx* gulper eels are small in relation to body size. They have a lens lying beneath a transparent "window" of skin. Although effective enough to make out flashes of bioluminescent light, the eyes cannot focus to produce a clear image.

Just in front of each eye lie two small nostrils that connect to an olfactory (smell) organ beneath them. In mature males, the olfactory organs swell in size. Males probably find mates by following a scent trail that leads them to a receptive female. In the featureless blackness of the deep ocean, female gulper eels probably release chemical attractants called pheromones to attract potential mates.

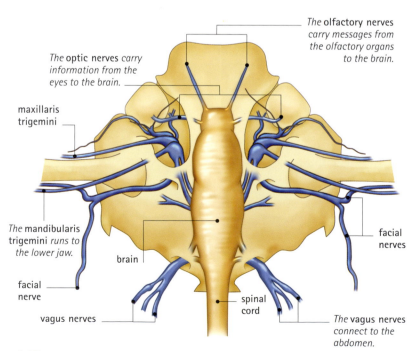

The optic nerves carry information from the eyes to the brain.

The olfactory nerves carry messages from the olfactory organs to the brain.

maxillaris trigemini

The mandibularis trigemini runs to the lower jaw.

brain

facial nerve

vagus nerves

spinal cord

facial nerves

The vagus nerves connect to the abdomen.

Two pores on either side of the head connect to the olfactory organ.

eye

lateral line organs

nasal lateral line

lateral line organs

Circulatory and respiratory systems

Like most other fish, gulper eels breathe using gills, where moving water is brought close to circulating blood, and gases are exchanged. Gulper eels live a sluggish life, and the midwaters of the deep ocean below 3,300 feet (1,000 m) have oxygen levels that remain reasonably high and more or less constant. For these reasons, the gills of the gulper eel do not need to be highly efficient to extract oxygen from the water. Gulper eel gills are plumelike, with only a few filaments. The gills lack the very large surface area found in more active fish. In *Saccopharynx* gulper eels there are four pairs of gill arches (rods of cartilage supporting the gills); in pelican eels there are five pairs. The gills need to be protected from damage by swallowed prey. The gill rakers of gulper eels are long and overlapping. They provide a protective barrier that prevents prey items—large or small—from entering the gill chamber.

Along with their other simplified organ systems, gulper eels do not need a highly efficient circulatory system. The heart and major blood vessels are based on the same plan as in other ray-finned fish, although they are less complex. The gulper eel's heart is small, and oxygen-carrying red blood cells typically make up less than 10 percent of the blood's volume. Fast-moving, shallow-water fish, such as mackerel, have a much higher demand for oxygen, and their blood contains as much as 50 percent red blood cells by volume.

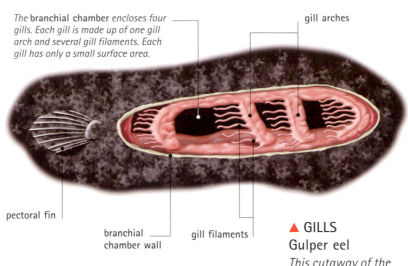

The **branchial chamber** encloses four gills. Each gill is made up of one gill arch and several gill filaments. Each gill has only a small surface area.

gill arches

pectoral fin

branchial chamber wall

gill filaments

▲ **GILLS**
Gulper eel
This cutaway of the body wall shows four gills in the gill chamber.

▼ *A pelican eel has five pairs of gill arches supporting the gills. The common Atlantic gulper eel has just four pairs.*

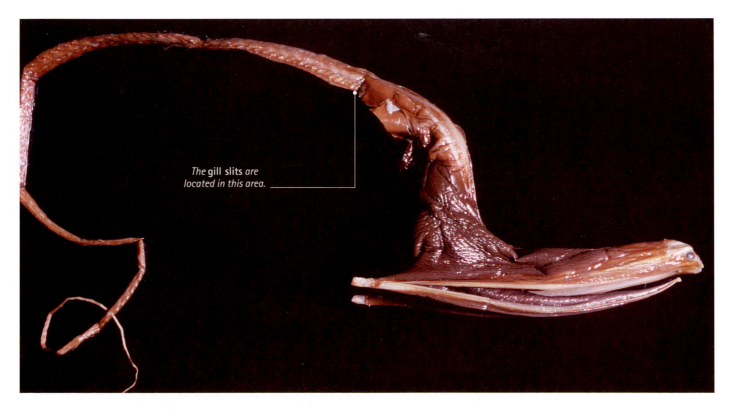

The **gill slits** are located in this area.

Digestive and excretory systems

The food of *Saccopharynx* gulper eels is fish prey that is typically very large in comparison with the size of the predator. In several of the specimens scientists have examined, the prey swallowed by the gulper eel weighed more than the gulper eel itself.

The gulper eel probably waves its tail's luminescent organ in front of its mouth to attract prey. When moved around, the filaments on the tail probably give the appearance of small swimming prey. Other fish come to investigate and are thus drawn to their doom. The very large jaw muscles of *Saccopharynx* gulper eels are vitally important when the fish are hunting. These muscles enable gulper eels to open their mouth very rapidly, creating a sudden decrease in pressure inside the mouth cavity. The fall in pressure sucks prey and water into the mouth.

Closely related pelican eels have tiny teeth compared with common Atlantic gulper eels and other *Saccopharynx* species. Pelican eels prey on much smaller creatures, especially shrimp. Unlike those of *Saccopharynx* gulpers, the body muscles of pelican eels suggest they

esophagus

*When the **mouth** opens suddenly, the partial vacuum inside draws in prey.*

*After feeding, the gulper eel's **stomach** is able to expand to several times its normal size.*

intestine

anus

▲ This deep-sea anglerfish has a bioluminescent lure projecting from its head. The lure attracts prey. Gulper eels probably use the bioluminescent "bulb" on their tail in much the same way.

▲ Common Atlantic gulper eel

The stomach cavity of many deep-sea fish is lined with the black pigment melanin. This dark barrier keeps swallowed bioluminescent animals from shining through the body wall, and so attracting unwanted attention from larger predators. Gulper eels, with their dark skin, do not require this additional light barrier.

can perform explosive bursts of speed for "sneak" attacks on prey. Pelican eels lack gulper eels' distensible, or stretchable, stomach.

Gulper eels lack pyloric ceca—fingerlike outgrowths of the gut just behind the stomach. In many predatory fish the ceca secrete digestive enzymes. Since the anus of a gulper eel would need to be enormously large to expel the leftover bones from a large meal, gulper eels sometimes expel their leftovers through the mouth. A gulper eel's kidneys are small and simplified.

Reproductive system

Female gulper eels produce eggs in their ovaries, and males manufacture sperm in testes. The eggs are very small, less than one-twelfth of an inch (2 mm) in diameter, and thousands are produced. As with most ray-finned fish, females probably release their eggs into the water, where they are fertilized by a male, but no one knows for sure. This method is called external fertilization.

When male and female gulper eels are ready to mate (they are "ripe"), they invest all their energy in finding each other and mating. In both sexes, the mature fish's search for a mate takes precedence over all other activities. Ripe gulpers no longer feed. They lose their teeth, and the lower jaw shrinks as energy is diverted to egg or sperm production. Other changes also occur, such as the loss of tail filaments. The eyes and olfactory organs of male gulper eels increase in size, presumably enabling them to better seek out females.

Since adult gulper eels move so little, the fish's larval phase (the leptocephalus larva) is the main dispersal phase in the life cycle. Once fertilized, the eggs float up into the surface waters. There, they hatch into larvae that feed on small plankton. Carried along on ocean currents, the growing larvae descend into the depths, maybe hundreds or even thousands of

IN FOCUS

Gulper eel larvae

Until the 1980s, some biologists believed that single-jaw eels, some individuals of which are only 4 inches (11 cm) or less in length, were the larvae of *Saccopharynx* gulper eels. Since then, adult single-jaw eels have been discovered, and specimens of *Saccopharynx* larvae found. The larvae live in the surface waters of the ocean and descend to greater depths as they mature. In gulper eel larvae, the myomeres, or muscle blocks, are V-shaped, not W-shaped as they are in shallow-water eel larvae.

▼ PREPARING TO BREED
Common Atlantic gulper eel
Several important changes occur in the body of a gulper eel when it is ready to mate. Internally, eggs and sperm are produced inside females and males, respectively. Externally, the eels lose their teeth, and males' eyes enlarge.

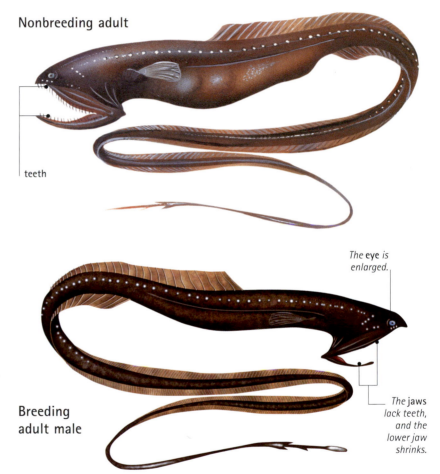

Nonbreeding adult

teeth

The **eye** *is enlarged.*

Breeding adult male

The **jaws** *lack teeth, and the lower jaw shrinks.*

miles away from where they hatched from their eggs. Like some shallow-water eels, gulper eels probably spawn once before dying.

TREVOR DAY

FURTHER READING AND RESEARCH
Moyle, P. B., and J. J. Cech. 2000. *Fishes: Introduction to Ichthyology*. Prentice Hall: Upper Saddle River, NJ.
Paxton, J. R., and W. N. Eschmeyer (eds). 1998. *Encyclopedia of Fishes*. Academic Press: San Diego, CA.
Randall, D. J., and A. P. Farrell (eds). 1997. *Deep-Sea Fishes*. Academic Press: San Diego, CA.

Hagfish

ORDER: Myxiniformes FAMILY: Myxinidae GENERA: *Eptatretus, Myxine, Nemamyxine, Neomyxine, and Notomyxine*

Hagfish are slimy, eel-like marine fish with a circular mouth that lacks jaws. The Atlantic hagfish is a typical species. It lives on and near the seabed of the North Atlantic and Arctic oceans at depths to about 3,150 feet (960 m). It feeds on bottom-living marine creatures such as shrimp and polychaete worms, and consumes dead and dying fish and other animals drifting down from the ocean's surface. Hagfish enter and devour their prey from the inside.

Anatomy and taxonomy

The Atlantic hagfish is one of 58 species of hagfish. Together with lampreys, they are usually classified as jawless fish, or agnathans, a group separate from all other vertebrates.

● **Animals** Hagfish, like other animals, are multicellular and gain their food supplies by consuming other organisms. Animals differ from other multicellular life-

forms in their ability to move from one place to another (in most cases, using muscles). They generally react rapidly to touch, light, and other stimuli.

● **Chordates** At some time in its life cycle a chordate has a stiff, dorsal (back) supporting rod called a notochord, which runs all or most of the length of the body. Hagfish have a notochord throughout their life.

● **Vertebrates** In all living vertebrates except hagfish, the notochord develops into a backbone (spine or vertebral column) made up of units called vertebrae. The vertebrate muscular system that moves the body consists primarily of muscles in a mirror-image arrangement on either side of the backbone or the notochord. That is called bilateral symmetry. Although usually classified among the vertebrates, hagfish do not have a backbone. However, they do have a simple cranium (braincase), another feature that is diagnostic of vertebrates.

● **Agnathans** The only living agnathans, or jawless fish, are the hagfish and lampreys. They have several features that make them distinct from all other living fishes—the gnathostomes, or jawed fish, such as the coelacanth, hammerhead shark, and trout. Features unique to

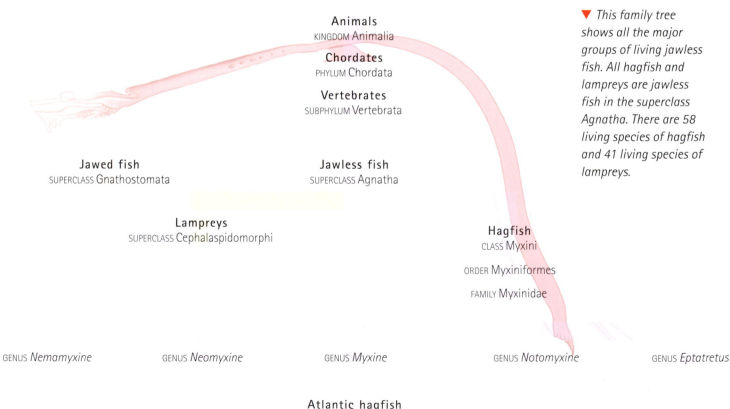

Animals
KINGDOM Animalia

Chordates
PHYLUM Chordata

Vertebrates
SUBPHYLUM Vertebrata

Jawed fish
SUPERCLASS Gnathostomata

Jawless fish
SUPERCLASS Agnatha

Lampreys
SUPERCLASS Cephalaspidomorphi

Hagfish
CLASS Myxini
ORDER Myxiniformes
FAMILY Myxinidae

GENUS *Nemamyxine* GENUS *Neomyxine* GENUS *Myxine* GENUS *Notomyxine* GENUS *Eptatretus*

Atlantic hagfish
GENUS AND SPECIES *Myxine glutinosa*

▼ *This family tree shows all the major groups of living jawless fish. All hagfish and lampreys are jawless fish in the superclass Agnatha. There are 58 living species of hagfish and 41 living species of lampreys.*

446

agnathans include a mouth that is not supported by jaws, gills (the breathing apparatus) that open to the outside through a pore or pores rather than slits, and the absence of pairs of fins.

Hagfish and lampreys look similar but are very different, both anatomically and physiologically. Both are eel-like, are slimy, and lack jaws, scales, and paired fins. However, lampreys have vertebrae, and hagfish do not. The blood of hagfish has a salt concentration similar to that of seawater; the blood of lampreys is much more dilute. Evidence from fossils and the comparative anatomy and biochemistry of living fish suggests that hagfish probably evolved at an earlier stage in the history of life than did lampreys. However, hagfish and lampreys have enough features in common to place them together as agnathans and separate from all other fish.

● **Lampreys** Lampreys are superficially similar to hagfish but differ from them in many anatomical and physiological features. Lampreys also differ in their mode of life and their life cycle. They belong to a single family, Petromyzontidae, and a single order, Petromyzontiformes, within the class Cephalaspidomorphi.

Of the 41 species of living lampreys, most are parasites or predators as adults. The adult's head is armed with a sucker on the ventral (under) side. The sucker bears numerous teeth that the lamprey uses to attach to its host or prey. A lamprey rasps a wound in the victim's body wall using a toothed tongue and then consumes the blood and tissues of its captive. The adult's single nostril is used solely for smell. A water current enters and leaves through seven pairs of gill openings. The head bears two eyes with a light-sensitive third eye, the pineal body, lying between them.

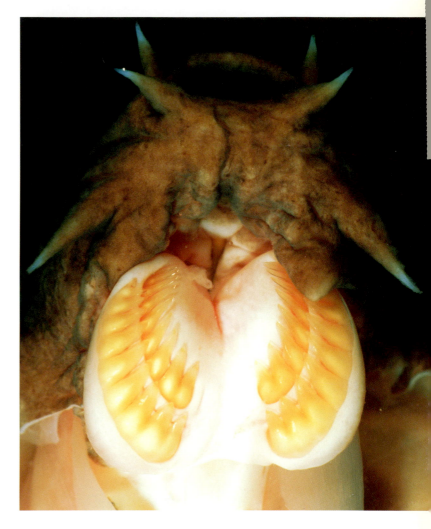

▲ *This hagfish has its dental plate everted (pushed out of its mouth). The dental plate is whitish, and the teeth are yellow. The funnel-shaped depression in the center is the mouth of the fish.*

EXTERNAL ANATOMY Hagfish are jawless fish with an eel-like slimy body, a tentacled head, and a caudal fin but no other fins. *See pages 449–451.*

SKELETAL AND MUSCULAR SYSTEMS The skeletal system is composed of cartilage, with a simple cranium, a framework that supports the gills, and a notochord that supports the body from head to tail. Blocks of muscle extend between the notochord and a system of inelastic fibers beneath the skin. Muscle blocks alternately contract and relax to flex the body, producing a wavelike swimming motion and an amazing knot-tying ability. *See page 452.*

NERVOUS SYSTEM The hagfish nervous system is a simple version of the normal vertebrate plan, with a brain and a spinal cord. The hagfish nervous system lacks myelin insulation around nerve fibers. *See pages 453–454.*

CIRCULATORY AND RESPIRATORY SYSTEMS The circulatory system is very unusual, with a four-chambered main heart, three accessory hearts, and large blood spaces, or sinuses, beneath the skin. The respiratory system is based on a series of gills inside unique structures called gill pouches. The hagfish also exchanges gases via the sinuses and across the skin. *See pages 455–456.*

DIGESTIVE AND EXCRETORY SYSTEMS The hagfish gut is simple and lacks a stomach. Food is digested inside a mucus bag that is finally expelled from the body wrapped around the feces. *See pages 457–458.*

REPRODUCTIVE SYSTEM Hatchling hagfish have the reproductive organs of both sexes, but usually develop as either male or female. The adult hagfish has a single gonad (egg- or sperm-producing organ). *See page 459.*

▲ *The very flexible notochord of a hagfish allows it to bend its body into a coil, as above, and even to tie itself into a tight knot. The latter ability helps the hagfish when it is ripping flesh from a dead animal.*

Lampreys have one or two dorsal (back) fins as well as a caudal (tail) fin. Adult lampreys have a backbone of vertebrae made of flexible cartilage.

Most lampreys are anadromous; this means that they hatch in freshwater and mature in the sea before returning to freshwater to mate and lay their eggs. Lampreys have a larval stage, the ammocoete, which gradually changes into a miniature version of the adult. Like hagfish, lampreys do not live in tropical or high polar regions.

● **Hagfish** The 58 described species of living hagfish are predators and scavengers. They belong to the single family Myxinidae and single order Myxiniformes within the class Myxini. An adult hagfish's mouth contains a tonguelike dental plate armed with teeth. The fish uses this apparatus to grasp food items and rip away chunks of flesh. The single nostril draws in water for respiration and to detect odors. A water current flows through a series of gill pouches and exits the body through gill pores. The gill pores may remain separate or merge into one gill opening. The subfamily Myxyninae includes those hagfish (in the genera *Myxine*, *Notomyxine*, *Neomyxine*, and *Nemamyxine* and including the Atlantic hagfish) that have a single gill opening on either side of the body. Members of the subfamily Eptatretinae (genus *Eptatretus*) have between 5 and 16 gill openings on either side. Adult hagfish have a cranium (skull) of fused cartilage and a notochord, but they lack a backbone with vertebrae.

A hagfish's head bears two skin-covered eyes. Three or four pairs of tentacles, or barbels, lie around the mouth and nostril; the tentacles are sensitive to touch and taste. Hagfish have a caudal fin but lack all other fins.

All hagfish live in temperate oceans. They inhabit the Atlantic, Indian, and Pacific oceans but do not enter warm tropical or cold polar regions, or freshwater.

External anatomy

CONNECTIONS

COMPARE the shape of a hagfish with that of a *SAILFISH* or a *TROUT*.

COMPARE the mouth of a hagfish with that of a *GULPER EEL*.

A hagfish looks like an eel armed with tentacles surrounding the mouth. The naked pink, purplish, or brown skin lacks scales and is extremely slimy. It is supplied with mucus by an array of between 70 and 200 glands. The glands empty onto the skin through a line of pores on either side of the body. A caudal, or tail, fin extends along part of the dorsal (back) and ventral (lower) surfaces of the body. Most hagfish are 12 to 24 inches (30–60 cm) long, but individuals of some species grow longer—for example, the napia.

Life without jaws

The hagfish's round mouth is ventrally placed and is shaped like a funnel. A dental plate on the tongue is equipped with horny teeth. A single tooth lies in the palate, which makes up the roof of the mouth. Several rows of teeth on either side of the dental plate fit together to rasp and scrape at food items. The action of the

EVOLUTION

Few fossils

There is a huge gap in the hagfish fossil record between the upper Carboniferous period (about 300 million years ago) and recent times. The single Carboniferous fossil hagfish is, however, extremely well preserved, and shows details of the head, jaws, and internal organs. This hagfish, called *Myxinikela siroka*, is strikingly similar to modern hagfish, suggesting that there has been little evolutionary change over that vast amount of time. Another fossil from slightly older rocks is called *Gilpichthys*. Some biologists tentatively classify it as a type of hagfish although it lacks the tentacles common to all other species.

▼ **Atlantic hagfish**
This hagfish has pinkish brown skin, several tentacles around its mouth and nostril, and a small caudal, or tail, fin.

*There are two pairs of paired **tentacles** around the nostril and two more around the mouth.*

mouth

*The hagfish draws water into its body through a single **nostril**.*

*After passing over the gills, water leaves the body through paired **gill openings**.*

*Mucus is released from **mucous gland pores**. The mucus gives the skin a slimy covering and deters many predators.*

*The **skin** of a hagfish is pink, brown, or purplish. It lacks scales.*

*The **caudal fin** extends along part of the dorsal and ventral surfaces of the body.*

*The **cloaca** is the joint sexual and excretory opening.*

32 inches (71–81 cm)

EVOLUTION

Are hagfish vertebrates?

Most fish biologists classify hagfish as vertebrates—members of the subphylum Vertebrata. However, there are strong grounds for placing hagfish within a subphylum of their own, the Myxini, within the phylum Chordata, which includes all the chordates. That hagfish are chordates is not in doubt; they have a notochord and many other chordate features, such as the organization of their nervous system, muscle blocks called myomeres, and a tail that lies behind the anus. Hagfish also have some vertebrate features, such as a cranium protecting the brain. However, a hagfish's notochord is not replaced by a vertebral column during the animal's development; hagfish are far from typical vertebrates.

dental plate teeth against one another and against the palate tears or rasps chunks of flesh from a victim.

Surrounding the mouth of a hagfish are paired sensory tentacles that sweep to and fro when the hagfish is hunting for food. A single nostril above the mouth draws water into the throat (pharynx) and then into gill pouches that contain gills. There, the oxygen that is needed for respiration is absorbed into the blood.

The main by-product of respiration, carbon dioxide, moves in the reverse direction. The

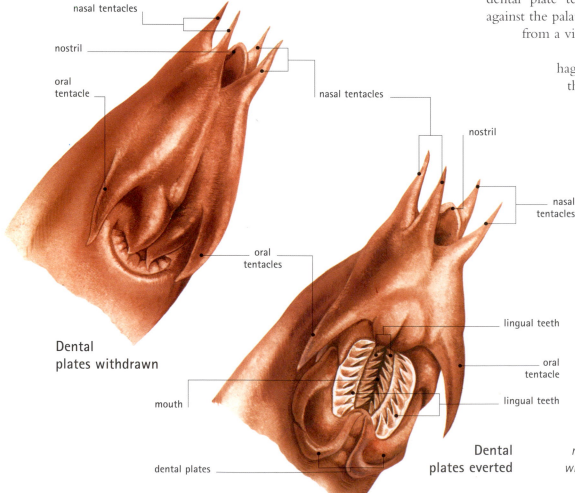

Dental plates withdrawn

- nasal tentacles
- nostril
- oral tentacle
- mouth
- dental plates

Dental plates everted

- nasal tentacles
- nostril
- nasal tentacles
- oral tentacles
- lingual teeth
- oral tentacle
- lingual teeth

◄ MOUTH AND HEAD
Hagfish
The hagfish turns its dental plates outward to expose the rasping banks of lingual teeth when it feeds. When a hagfish is not feeding, its lingual teeth are withdrawn inside its mouth.

hagfish nostril is lined with sensory cells that detect chemicals dissolved in the water.

How water moves out

After passing across the gills, water leaves the gill pouch through a gill pore. In most cases, gill pores merge to exit the body through a reduced number of gill openings on either side of the body. There is one gill opening in most groups, but there are between 5 and 16 in the *Eptatretus* hagfish.

In all hagfish species one or two larger openings occur on the left flank behind the gill openings. Any items that would otherwise block the respiratory system are expelled from the body through these openings.

Poor eyesight

Hagfish eyes lie beneath the skin. They can detect differences in light intensity but lack lenses and do not focus images. The poor eyesight of hagfish is not surprising, since they live mostly in the murky ocean depths, where there is little light. Hagfish rely less on vision and more on smell, taste, and touch to detect prey, predators, and potential mates.

▲ *The small pale spots on this hagfish are mucous pores. Slimy mucus from glands is released through the pores to give the hagfish a slippery protective covering. The larger pale areas are gill openings.*

Spectacular slime

A row of glands on the sides of a hagfish release slime through pores. The slime is a mixture of mucus and protein threads. The protein threads extend swiftly on contact with seawater to trap the mucus around the skin. The mucus sticks to almost any object, and is a powerful deterrent to predators. If a hagfish is placed in a bucket of seawater for a few minutes, its slime will change the consistency of the water to that of wallpaper paste. The slime makes the hagfish almost impossible to grasp, clogs the gills of attacking fish, and is probably unpalatable to most predators. Hagfish even coat their food with slime to keep other scavengers from stealing it. When the hagfish's slime covering threatens to block its own gill openings, it ties a knot in its tail. The fish slides the knot along its body, pushing the excess slime away. The hagfish then exhales water through its nostril to "sneeze" away any blocking mucus.

Skeletal and muscular systems

COMPARE the simple skeleton and fin of the hagfish with the arrangement of skeleton and fins in a cartilaginous fish such as a *HAMMERHEAD SHARK* or in a bony fish such as a *TROUT*.

CONNECTIONS

The skull, or cranium, of a hagfish is a simple structure made of cartilage. Most sections of the cranium are fused. A branchial skeleton of cartilage provides some support to the gills. A hagfish does not have a backbone but retains a cartilaginous (composed of cartilage) notochord throughout its life. There is no evidence, either in the fossil record or from examining the developmental stages of hagfish, to suggest that these animals have ever had a recognizable backbone.

In the absence of the usual vertebrate skeleton with a backbone and ribs, the dermal layer beneath the skin not only provides a tough envelope that maintains body shape but also offers a framework to which body muscles can attach. The dermis is rich in collagen fibers that are resistant to stretching. However, the fibers are so arranged that they allow the body to flex for swimming and knot tying.

The skeleton of a lamprey, which is related to hagfish, is slightly more complex. A lamprey's notochord has several pairs of cartilaginous projections, called arcualia, extending from it, and the skull has a more complex structure.

As in other fish, the blocks of muscle used in swimming, called myomeres, are separated by

IN FOCUS

Tied in knots

The notochord supports the hagfish's body, from the base of the cranium to the tail. The notochord's flexibility enables the hagfish to tie its body in a knot, an ability the fish employs in consuming food, wiping mucus from its body, and wriggling from the grasp of predators.

partitions of connective tissue. Blocks on opposite sides of the notochord contract antagonistically (against one another) causing the body to bend first one way and then the other. This action produces S-shaped waves that move backward along the body, pushing against the water and driving the hagfish forward. Many of the muscle cells in a hagfish are able to work at low concentrations of oxygen. This ability is important because hagfish are burrowing animals that often bore inside their food, so they often rest and feed in oxygen-poor conditions.

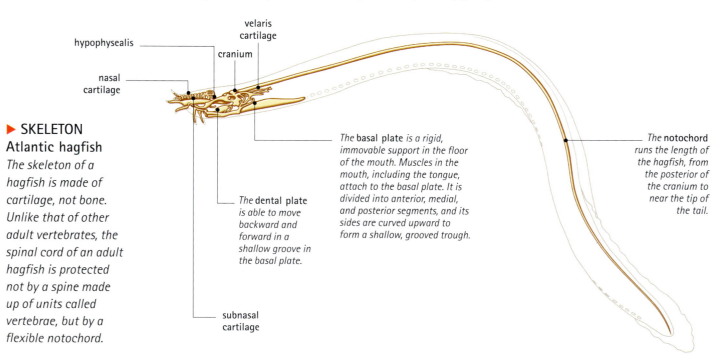

▶ **SKELETON Atlantic hagfish**
The skeleton of a hagfish is made of cartilage, not bone. Unlike that of other adult vertebrates, the spinal cord of an adult hagfish is protected not by a spine made up of units called vertebrae, but by a flexible notochord.

hypophysealis

nasal cartilage

velaris cartilage

cranium

*The **basal plate** is a rigid, immovable support in the floor of the mouth. Muscles in the mouth, including the tongue, attach to the basal plate. It is divided into anterior, medial, and posterior segments, and its sides are curved upward to form a shallow, grooved trough.*

*The **dental plate** is able to move backward and forward in a shallow groove in the basal plate.*

subnasal cartilage

*The **notochord** runs the length of the hagfish, from the posterior of the cranium to near the tip of the tail.*

Nervous system

Compared with that of more typical vertebrates, the nervous system of a hagfish is primitive. As in other vertebrates, the central nervous system (CNS) consists of the brain and spinal cord. The CNS is connected via the nerves of the peripheral nervous system (PNS) to sensory organs and to responsive structures, such as muscles.

The spinal cord and brain

The hagfish spinal cord is flattened and similar in appearance to that of lampreys, although the lamprey's is thicker. The hagfish brain is broadly divided into three regions, as in other vertebrates: the fore-, hind-, and midbrain. In detailed structure, however, the hagfish brain is unlike that of any other vertebrate except the lamprey. The forebrain has an unusually large region devoted to smell (the olfactory bulb). The midbrain, as in other vertebrates, is a relay center for directing nerve impulses from sensory organs to parts of the forebrain. The hagfish hindbrain contains the medulla, which coordinates automatic activities such as breathing, but—unusually—not the beating action of the various hearts. Heartbeats are controlled by hormones (chemical messengers

released into the blood) rather than by direct nervous action. The hindbrain of hagfish and lampreys has only a poorly developed cerebellum, the region that, in other vertebrates, is responsible for the coordination of balance and body movement.

No insulation

In hagfish, the axon (elongated region) of nerve cells lacks a fatty insulating layer called myelin. In other vertebrates, the myelin sheath

▼ BRAIN
Atlantic hagfish
The brain is around 0.2 inch (0.5 cm) long. Two optic nerves connect the eyes with the underside of the forebrain.

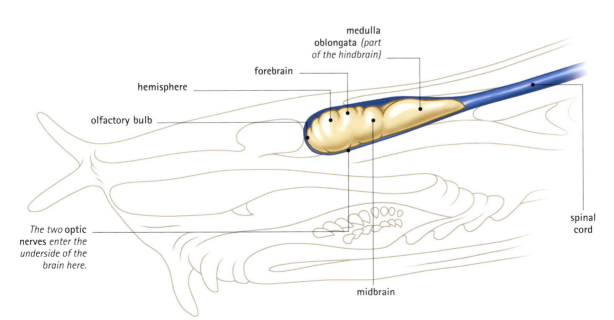

medulla oblongata *(part of the hindbrain)*

forebrain

hemisphere

olfactory bulb

*The two **optic nerves** enter the underside of the brain here.*

spinal cord

midbrain

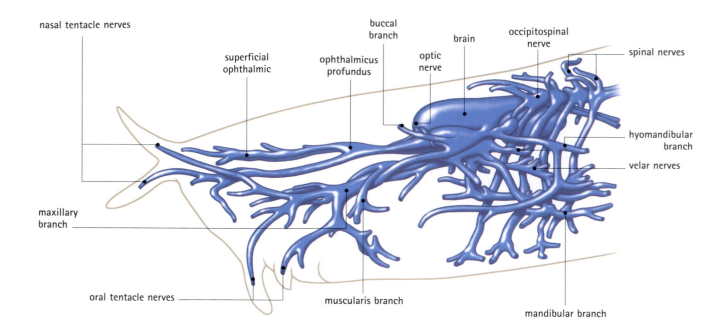

nasal tentacle nerves

superficial
ophthalmic

ophthalmicus
profundus

buccal
branch

optic
nerve

brain

occipitospinal
nerve

spinal nerves

hyomandibular
branch

velar nerves

maxillary
branch

oral tentacle nerves

muscularis branch

mandibular branch

▲ **CRANIAL NERVES**
Hagfish

The major nerves of the hagfish's head are the ophthalmic, maxillary, and mandibular nerves. Together they are sometimes called the trigeminal nerves.

insulates nerve fibers against electrical interference from other fibers. The sheath also increases the rate at which nerve impulses can travel through the nervous system, since it contains gaps that serve to speed the signal. So, nerve conduction in the hagfish is unusually slow. Some activities, such as heartbeat, that are controlled by the nervous system in other vertebrates are regulated by blood-circulated hormones in hagfish.

Hagfish senses

The hagfish has a balance organ, a semicircular canal, inside the head. Scientists once believed

that all hagfish lacked a lateral-line system, the vibration-sensing series of cells that lie along the flanks of other fish. However, some species of hagfish do have traces of such a system. Biologists do not know whether this is similar to an earlier evolutionary stage in other fish, or whether it was once a more sophisticated system that has become simpler over millions of years of evolution.

The head of a hagfish, and in particular the tentacles, is packed with both touch-sensitive and chemical-detecting cells. The single nostril is an opening above the mouth that leads to the pharynx and a blind-ended smell sac. Inside the smell sac are chemical-detecting cells that provide the hagfish's sense of smell. Sensory organs are scattered densely over the hagfish's snout and tentacles, and are distributed more sparsely over much of the rest of the body. These organs resemble the taste buds of other vertebrates and "taste" the water.

The sensory world of the hagfish is one of touch, taste, and smell. Sight is less important. Present-day hagfish have poor vision. Living on a muddy or sandy seabed in deep water where little or no light is present, they do not need good eyesight, since their other senses can compensate. In *Myxine*, portions of skin distant from the eyes, such as a region near the cloaca (the common opening into which the reproductive, digestive, and excretory systems empty), are also sensitive to light.

EVOLUTION

Lancelets

Lancelets are fishlike creatures that filter food from the water using a set of fine bristles around their mouth. The animals live on shallow, sandy seabeds. Lancelets are invertebrate chordates that have a mix of vertebrate and invertebrate features, which hint at the sort of body organization that occurred during the early stages of vertebrate evolution. Lancelets have a notochord, a dorsal nerve cord (a type of spinal cord), and a gut with a post-anal tail—all typical vertebrate features. However, the way different body layers develop, the structure of their sperm, and the connections between muscle blocks and the nerve cord are more typical of echinoderms such as starfish than of vertebrates. For example, the muscle blocks in a lancelet are connected to the nerve cord by "muscle tails" rather than by peripheral nerves, as occurs in vertebrates.

Circulatory and respiratory systems

CONNECTIONS

COMPARE the position of a hagfish's gills with those of a *TROUT*.

COMPARE the structure of a hagfish's heart with the that of an aquatic mammal, such as a *DOLPHIN* or a *GRAY WHALE*.

Hagfish inhale water not through their mouth but through their single nostril. A current of water flows from the nostril to the gills. The current is created by folds of tissue called velar folds. They roll and unroll inside the pharynx, or throat, of the hagfish. At the same time, muscles in the walls of gill pouches (which enclose the gills), and in the gill pores that lead to the outside, squeeze from front to back. This movement further encourages the flow of water over the gills. The breathing mechanism comprising velar folds and gill pouches is unique to hagfish.

A countercurrent system

In some species of hagfish the gill pores on one side of the body merge into a common gill opening. In *Eptatretus* hagfish the gill pores remain separate or merge into small groups; water exits the body through five or more gill openings on the side of the body. Inside the gills, water flows in the direction opposite to the flow of blood. This counterflow system is similar to that of many other species of fish. Blood leaving the gill meets newly arriving water that is rich in oxygen and low in carbon dioxide. Such a countercurrent arrangement maximizes the rates of oxygen absorption and carbon dioxide excretion.

Hagfish can gain the oxygen they need, and expel carbon dioxide, even when the head is buried deep in food. This means that their gill function must be supplemented by other means of gas exchange. Large blood-filled spaces called sinuses lie just beneath the skin, and gas exchange can occur through their walls. Hagfish also absorb oxygen and excrete carbon dioxide through their skin. Hagfish skin is rather like a loose-fitting sock; if the fish is held aloft by the head, blood settles and swells the tail region.

A hagfish has about twice the volume of blood of a jawed fish of similar size. Hagfish have a low rate of metabolism (the overall rate of chemical reactions in the body) so their demand for oxygen is relatively low. Breathing through the skin can usually satisfy their demand for oxygen except when they are very active. The main heart, the accessory hearts, and many other tissues are able to gain energy from the breakdown of food substances even when oxygen levels are low.

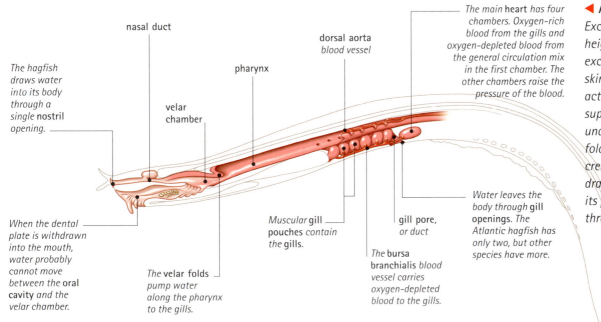

nasal duct

dorsal aorta blood vessel

The main **heart** has four chambers. Oxygen-rich blood from the gills and oxygen-depleted blood from the general circulation mix in the first chamber. The other chambers raise the pressure of the blood.

pharynx

velar chamber

The hagfish draws water into its body through a single **nostril** *opening.*

When the dental plate is withdrawn into the mouth, water probably cannot move between the **oral cavity** *and the velar chamber.*

The **velar folds** *pump water along the pharynx to the gills.*

Muscular gill pouches *contain the* **gills.**

The **bursa branchialis** *blood vessel carries oxygen-depleted blood to the gills.*

gill pore, *or duct*

Water leaves the body through **gill openings.** *The Atlantic hagfish has only two, but other species have more.*

◄ **Atlantic hagfish**
Except during periods of heightened activity, gas exchange through the skin is sufficient. When active a hagfish can supplement this by undulating its velar folds. This action creates a current, drawing water along its pharynx and through its gills.

▶ MAIN HEART
Atlantic hagfish

The heart pumps blood to the gills and around the body. Oxygen-depleted blood passes through the four chambers of the heart. Blood pressure is raised as it is pumped through. The blood then passes into tiny blood vessels in the gill pouches. There, gas exchange occurs. Oxygen passes into red blood cells in the bloodstream, and carbon dioxide is removed from the blood.

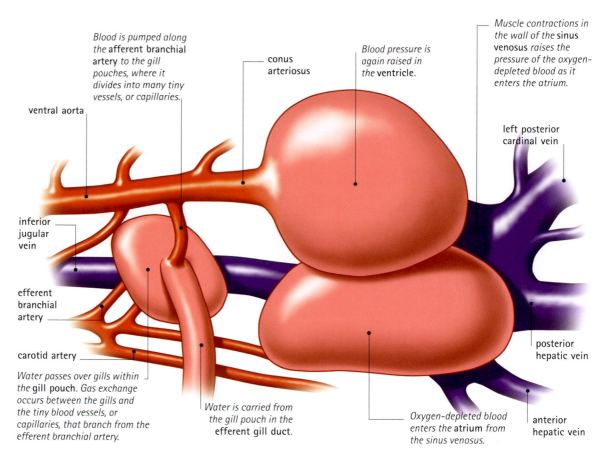

*Blood is pumped along the **afferent branchial artery** to the gill pouches, where it divides into many tiny vessels, or capillaries.*

conus arteriosus

*Blood pressure is again raised in the **ventricle**.*

*Muscle contractions in the wall of the **sinus venosus** raises the pressure of the oxygen-depleted blood as it enters the atrium.*

ventral aorta

left posterior cardinal vein

inferior jugular vein

efferent branchial artery

carotid artery

*Water passes over gills within the **gill pouch**. Gas exchange occurs between the gills and the tiny blood vessels, or capillaries, that branch from the efferent branchial artery.*

Water is carried from the gill pouch in the efferent gill duct.

*Oxygen-depleted blood enters the **atrium** from the sinus venosus.*

posterior hepatic vein

anterior hepatic vein

Hagfish hearts

The main, or brachial, heart of a hagfish has four chambers, with the chambers arranged in a line. Oxygen-rich blood from the gills and oxygen-depleted blood from the rest of the body mix in the first heart chamber, the sinus venosus. Weak contractions in the wall of the sinus venosus deliver blood to the atrium, which in turn pumps blood into the thick-walled ventricle. These three chambers, one after the other, gradually raise the pressure of the blood until it is high enough to be pumped through the tiny blood vessels of the gills. The fourth chamber, called the conus arteriosus, has elastic walls that moderate the extremes of pressure and help ensure that the blood is delivered more or less continuously, rather than in a series of pulses.

A hagfish also has several other heartlike structures in its circulatory system in addition to the main heart. Two of these accessory hearts (the cardinal and caudal hearts) raise the blood's pressure after it has passed through the gills. Another accessory heart, called the portal heart, raises the pressure of blood before it passes through the liver. Other structures raise the blood pressure at the exits of the sinuses. The sinuses lie directly under the skin and are a unique feature of hagfish. Despite the pumping action of the accessory hearts, the presence of the sinuses causes hagfish to have the lowest blood pressure of any vertebrate.

CLOSE-UP

Red blood cells

In hagfish and lampreys, the blood's oxygen-carrying pigment, hemoglobin, is made up of a single polypeptide chain (a chain of amino acid molecules) rather than the four polypeptide chains found in all other vertebrates. Hagfish hemoglobin is not as efficient as many other fish hemoglobins at transporting oxygen. However, the oxygen-carrying capacity of hagfish hemoglobin is sufficient for these sluggish animals to live even in the low-oxygen conditions in which they often are found.

Digestive and excretory systems

Although hagfish lack jaws, they have a flexible dental plate. The plate is armed with horny teeth made of a tough protein called keratin (which forms fingernails in humans). During rest, the plate is folded inside the mouth so the teeth point inward. During feeding, the plate swings forward, and the teeth can grasp objects. By moving the plate back and forth, in and out of the mouth, the hagfish can rasp at food.

Most species of hagfish feed on bottom-living invertebrates, particularly polychaete worms and crustaceans such as shrimp. Hagfish also scavenge on dead and dying fish drifting down from surface waters. They home in on their food using chemical detection; they may also detect the vibrations of an ailing fish during its death throes, or the disturbance caused by the attention of other scavengers. When the prey is found, the hagfish usually enters its body through an existing entrance, such as the mouth or a gill slit. Alternatively, it may use its sharp dental plate to rip a hole in the body wall of the fish. Usually the hagfish consumes the fish from the inside out. It is not uncommon for trawler fishers to find the skin and skeleton of a captured fish with a well-fed hagfish inside. While feeding, a hagfish

sometimes grasps a carcass with its mouth and then passes a knot along the length of its body, from tail to head. This levers the hagfish's head away from the carcass, taking a chunk of the victim's flesh with it.

Bags of mucus

The hagfish has a tubular gut that lacks a proper stomach for temporarily storing food. Once swallowed, the hagfish's food is wrapped in a bag of mucus secreted by the gut wall. Digestive enzymes pass into the mucus bag, and digested food passes out. The bag is

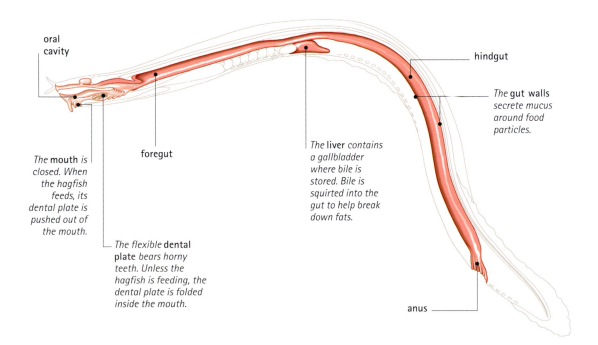

oral cavity

hindgut

The **gut** walls secrete mucus around food particles.

The **mouth** is closed. When the hagfish feeds, its dental plate is pushed out of the mouth.

foregut

The **liver** contains a gallbladder where bile is stored. Bile is squirted into the gut to help break down fats.

The flexible **dental plate** bears horny teeth. Unless the hagfish is feeding, the dental plate is folded inside the mouth.

anus

◀ **Atlantic hagfish**
With the mouth closed, the oral cavity of this hagfish is in front of the dental plate. When it feeds, food passes from the oral cavity into the gut, where digestion occurs. Waste products are ejected through the anus.

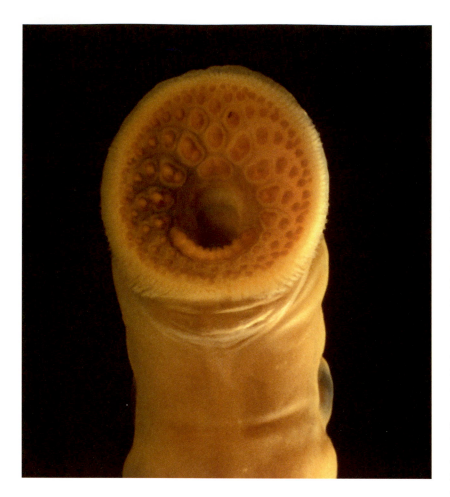

▲ *This is a lamprey, a relative of the hagfish. A sucker surrounds the central mouth depression, and the lamprey uses this to attach itself to prey. A tonguelike apparatus with tough teeth emerges from the mouth to rasp at flesh.*

▶ **FEEDING KNOT**
Atlantic hagfish
After securing its grip on a dead fish, the hagfish throws its body into a knot. The hagfish slides the knot down to its head, levering its mouth away from its prize and tearing a chunk of flesh away.

eventually expelled as a neat wrapper around the solid waste. Their metabolism, or rate of chemical reactions in the body, is comparatively slow, and hagfish are able to survive without feeding for months on end.

The liver of a hagfish is surprisingly large, and like that of most other vertebrates it contains a gallbladder where bile is stored. Bile is squirted into the gut to emulsify fats, breaking them down into small droplets for easier digestion. The structure and output of

the hagfish liver differ strikingly from those of other vertebrates. Most vertebrates have a lobe-shaped liver with cells arranged around blood vessels. A hagfish liver is tubular. The bile of most vertebrates contains bile salts. Uniquely, hagfish bile contains alcohol.

The concentration of salts in the tissues of hagfish is similar to that of seawater. For this reason, many biologists suspect that hagfish evolved in the oceans and have always lived there. However, some features of the hagfish kidney hint that this theory might not fit all the facts. Experiments with *Eptatretus* hagfish show that if the sodium content of the surrounding seawater is lowered, the hagfish kidney responds by absorbing more sodium into the bloodstream and excreting less in the urine. The ability to actively absorb sodium is not necessary in seawater, where sodium levels are high and remain more or less constant, but is essential for freshwater vertebrates as a means of conserving the chemical.

Although the hagfish kidney has a simple structure, it has features found in other fish that would enable the hagfish to dilute its urine. This ability is vital in freshwater, where an animal needs to retain its salts rather than lose them in urine, but it is unnecessary in the oceans. This evidence suggests that hagfish had a freshwater phase in their early evolutionary history. If so, they may have been anadromous—they spawned in freshwater but matured in the sea— as are many lampreys today.

Water diffuses relatively freely back and forth through the hagfish's skin, but since the salt concentration inside the body is similar to that outside there is no overall movement of water in or out across the body wall. Under normal conditions, a hagfish produces little urine.

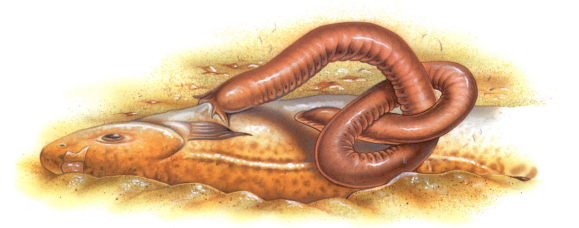

Reproductive system

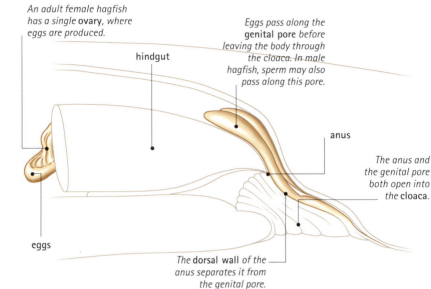

An adult female hagfish has a single **ovary**, where eggs are produced.

Eggs pass along the **genital pore** before leaving the body through the cloaca. In male hagfish, sperm may also pass along this pore.

hindgut

anus

The anus and the genital pore both open into the **cloaca**.

eggs

The **dorsal wall** of the anus separates it from the genital pore.

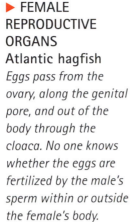

▶ FEMALE REPRODUCTIVE ORGANS
Atlantic hagfish
Eggs pass from the ovary, along the genital pore, and out of the body through the cloaca. No one knows whether the eggs are fertilized by the male's sperm within or outside the female's body.

▼ EGGS
Atlantic hagfish
Female hagfish lay 20 to 30 leathery eggs that are about 1 inch (2.5 cm) long. The eggs have long filaments tipped with hooks on one of their ends. They anchor the eggs to the seabed and often link the eggs together in rosette-shaped clusters.

Biologists know little about how hagfish reproduce. During the 1860s a Danish scientific organization offered a prize to anyone who could describe the early development of the life cycle of the European hagfish. The prize went unclaimed and was withdrawn decades later. It is very rare to find live hagfish eggs, and young hagfish have never been caught or observed in the wild.

Single sex organs

Both hagfish and lampreys have a single gonad (sperm- or egg-producing organ). The gonad sheds sex cells (eggs or sperm) into the fluid-filled body cavity, or coelom. This is unusual; other vertebrates shed their sex cells directly into the reproductive ducts. In hagfish and lampreys, the sex cells, or gametes, eventually pass from the coelom through a pair of genital pores that join ducts leading from the kidney. The eggs and sperm exit the body through the cloaca, an opening for the release of sex cells as well as solid waste and urine.

Male hagfish do not have obvious external reproductive organs, and it is not clear how they fertilize the females' eggs. Despite the presence of a tough outer case in most hagfish eggs, it is possible that males shed their sperm directly onto newly laid eggs before they harden. Alternatively, males may fertilize the eggs internally by mating with the female

before the eggs are laid. The gonad of a male hagfish is small, suggesting that relatively little sperm is produced. Animals with small gonads usually practice internal fertilization, but zoologists do not know for certain how fertilization occurs in these fish. Hagfish eggs hatch after two months or more, and juveniles emerge as miniature versions of the adults. Unlike lampreys, hagfish have no larval phase.

TREVOR DAY

FURTHER READING AND RESEARCH
Moyle, P. B. and J. J. Cech. 2000. *Fishes: Introduction to Ichthyology.* Prentice Hall: Upper Saddle River, NJ.
Paxton, J. R. and W. N. Eschmeyer (eds). 1998. *Encyclopedia of Fishes.* Academic Press: San Diego, CA.

COMPARATIVE ANATOMY

Strange sex-shifters

Young hagfish possess the reproductive organs of both sexes. However, usually only one set of organs matures to determine the sex of the adult. In a few species, adults may retain the ability to change sex. They may be male one season and female the next. Such sex-switching is surprisingly common in the animal world. Many fish, shrimp, worms, and mollusks change sex as they age. Strangely, in such species the switch almost always occurs when the animal has reached 72 percent of its maximum size. Biologists cannot explain the significance of this figure.

Hammerhead shark

ORDER: Carcharhiniformes FAMILY: Sphyrnidae
GENUS: Sphyrna

The nine species of hammerhead sharks are marine fish that live in tropical and warm-temperate oceans all around the world. A few species, such as the smooth hammerhead, may stray into cooler waters, but that is rare. They hunt mainly in shallow coastal seas.

Anatomy and taxonomy

Scientists group all organisms into taxonomic groups based largely on anatomical features. Hammerhead sharks belong to the order Carcharhiniformes, the largest of several orders in the subclass Elasmobranchii, which includes all the sharks and rays.

● **Animals** Animals are multicellular (many-celled) organisms with well-developed powers of movement, using muscles, and the ability to respond rapidly to stimuli. They typically obtain nutrients by eating other organisms. They digest the tissues of these organisms, breaking them down into simple molecules. These are used to provide energy for the animal or to build new tissues.

● **Chordates** A chordate has a strong but flexible rod called a notochord extending along its back. This rod supports the animal's body and provides places of attachment for many of its muscles. Some chordates lose the notochord as they reach adulthood, but some keep it throughout their lives.

● **Vertebrates** The notochord of a vertebrate forms the basis of a flexible backbone (also called a spine or vertebral column), which is composed of units called vertebrae. A vertebrate's brain is enclosed within a cranium, or braincase. The vertebrate muscular system that moves the head, trunk, and limbs consists primarily of muscles that are bilaterally symmetrical about the skeletal axis. In other words, the muscles on one side of the backbone are the mirror image of those that occur on the other side.

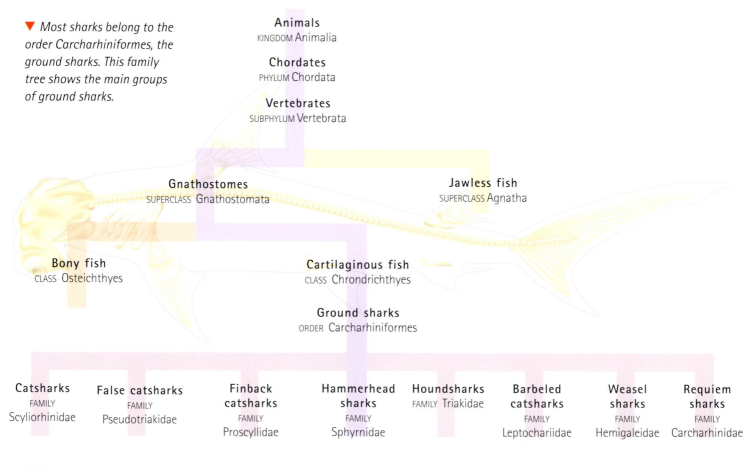

▼ Most sharks belong to the order Carcharhiniformes, the ground sharks. This family tree shows the main groups of ground sharks.

Animals
KINGDOM Animalia

Chordates
PHYLUM Chordata

Vertebrates
SUBPHYLUM Vertebrata

Gnathostomes
SUPERCLASS Gnathostomata

Jawless fish
SUPERCLASS Agnatha

Bony fish
CLASS Osteichthyes

Cartilaginous fish
CLASS Chrondrichthyes

Ground sharks
ORDER Carcharhiniformes

Catsharks
FAMILY Scyliorhinidae

False catsharks
FAMILY Pseudotriakidae

Finback catsharks
FAMILY Proscyllidae

Hammerhead sharks
FAMILY Sphyrnidae

Houndsharks
FAMILY Triakidae

Barbeled catsharks
FAMILY Leptochariidae

Weasel sharks
FAMILY Hemigaleidae

Requiem sharks
FAMILY Carcharhinidae

● **Cartilaginous fish** Many vertebrates, such as birds and mammals, have a skeleton made of hard bone, but the backbone, skull, and other skeletal units of a cartilaginous fish are made of softer, more flexible cartilage. These fish also have enamel-covered "placoid" skin scales, or denticles, which have the same basic structure as the teeth.

● **Sharks and rays** Sharks are mainly streamlined, active predators with sharp teeth and very acute senses, although some types have flat teeth and feed on shellfish. A shark's teeth develop in rows, with each row gradually moving forward to replace teeth that are damaged or lost. Rays are adapted for feeding on the sea bottom and have a flat,

▲ *The great hammerhead shark usually grows to a length of around 13 feet (4 m) but may grow to as much as 20 feet (6 m).*

camouflaged body. Fertilization in both sharks and rays is internal, the male transferring sperm into the female's body using a pair of "claspers."

● **Ground sharks** The largest order of sharks, the Carcharhiniformes, include a wide range of types, from inactive, small-toothed species to powerful hunters such as the requiem and hammerhead sharks. In common with most other sharks, ground sharks have an asymmetrical tail, with the upper lobe larger than the lower lobe.

FEATURED SYSTEMS

EXTERNAL ANATOMY Sleek and streamlined, hammerhead sharks are named for the unique shape of their head. *See pages 462–463.*

SKELETAL SYSTEM The skeleton is made not of bone, but from a softer, more flexible material called cartilage. The sharp-toothed jaws are not rigidly attached to the skull. *See pages 464–465.*

MUSCULAR SYSTEM Big flank muscles provide great power for fast pursuit swimming, while other muscles are adapted for endurance swimming. *See pages 466–467.*

NERVOUS SYSTEM Highly developed senses enable hammerhead sharks to detect prey from great distances and target it with perfect precision, even in the dark. *See pages 468–471.*

CIRCULATORY AND RESPIRATORY SYSTEMS Hammerhead sharks gather oxygen from water flowing over blood-filled gills. Blood is pumped around the body partly by the heart and partly by the action of the swimming muscles. *See pages 472–473.*

DIGESTIVE AND EXCRETORY SYSTEMS Hammerhead sharks have an elastic stomach, allowing them to eat big meals. Food passes slowly through a spiral digestive structure to increase the efficiency of food breakdown. The liver is large and filled with buoyant oil that helps stop the shark from sinking. *See page 474.*

REPRODUCTIVE SYSTEM Fertilization of eggs is internal, and a female hammerhead gives birth to live, well-developed young. *See page 475.*

External anatomy

CONNECTIONS

COMPARE the dorsal fin of a hammerhead shark with the dorsal fin of a *SAILFISH*. The sailfish's fin is a thin membrane supported by strong bony struts.

COMPARE the body form of a shark with that of the *STINGRAY*. The shark is adapted for hunting in open water, while the flattened ray hunts on the seafloor.

In many ways a hammerhead is a typical large shark. It has a sleek, muscular, streamlined body, ideally shaped for slipping through the water with minimum effort. Its big tail, with a tall upper lobe, provides plenty of thrust. Its long pectoral fins provide stability and lift. The pectorals have the same shape in cross section as the wings of an airplane, and they do much the same job but in water rather than air. The dorsal fin is tall and pointed just like that of any of the requiem sharks—a large family that also includes species like the gray reef shark. However, as their name suggests, hammerheads have a unique and unusual head structure.

Instead of a pointed snout, a hammerhead shark head looks more like a wing, with an eye and nostril at each "wing tip." This shape is called a cephalofoil. In one species, the winged hammerhead shark, the width of the cephalofoil can be equal to half the length of its body.

In cross section the cephalofoil is shaped like a fin. This helps it move through the water while minimizing drag (the force produced by water resistance). The cephalofoil may even generate some lift to aid stability. However, its main function is to accommodate a powerful system of electric field detectors.

Reinforced skin

A hammerhead shark does not have a scaly body like most fish. It has tough, coarse skin reinforced with cross-fibers that work like the woven fabric backing of a rubber tire. Its skin surface is covered with small toothlike studs

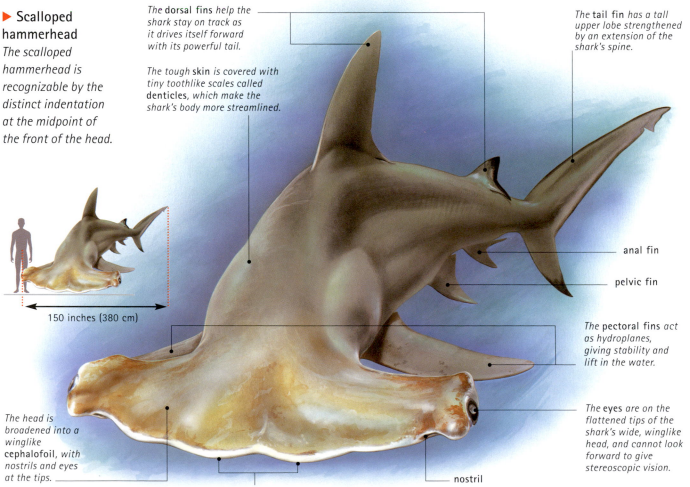

▶ **Scalloped hammerhead**
The scalloped hammerhead is recognizable by the distinct indentation at the midpoint of the front of the head.

The **dorsal fins** help the shark stay on track as it drives itself forward with its powerful tail.

The tough **skin** is covered with tiny toothlike scales called **denticles**, which make the shark's body more streamlined.

The **tail fin** has a tall upper lobe strengthened by an extension of the shark's spine.

150 inches (380 cm)

anal fin

pelvic fin

The **pectoral fins** act as hydroplanes, giving stability and lift in the water.

The head is broadened into a winglike **cephalofoil**, with nostrils and eyes at the tips.

The **eyes** are on the flattened tips of the shark's wide, winglike head, and cannot look forward to give stereoscopic vision.

scallop

nostril

called placoid scales or denticles. All sharks and rays have denticles, and in species such as the bramble shark the denticles take the form of big, sharp spines.

A hammerhead's fins are fleshy and cannot be folded close to the shark's body like the fanlike, and often semitransparent, fins of a bony fish. The tail end of most sharks sweeps up to form the upper lobe of the tail fin.

COMPARATIVE ANATOMY

Slow sharks

People usually think of sharks as sleek, fast-moving hunters, but many are not like that at all. For example, angel sharks have a horizontally flattened body adapted for lying on the seabed, often half-buried in the sand, where their camouflage enables them to ambush fish, crabs, and other animals. The slow-moving, nocturnal cat sharks are similar, though they often prefer to scavenge for scraps rather than hunt live prey. One of the strangest sharks is the eel-like frilled shark. It has an elongated body adapted for hunting in rock crevices and caves.

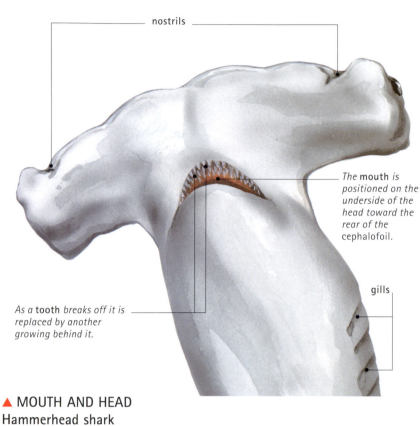

nostrils

The **mouth** is positioned on the underside of the head toward the rear of the cephalofoil.

gills

As a **tooth** breaks off it is replaced by another growing behind it.

▲ MOUTH AND HEAD
Hammerhead shark
The hammerhead shark's mouth is toward the rear of the head's underside. The widely spaced nostrils help the shark detect the direction from which an odor is coming.

IN FOCUS

Toothlike scales

The denticles of sharks have exactly the same structure as the teeth; a base of dentine is covered by hard, shiny enamel-like substances. The denticles form a hard armor that protects the shark's body. They may also enhance streamlining by improving water flow over the skin. However, if a shark's skin is rubbed the "wrong" way, from tail to head, the sharp denticles make it feel very rough—in the past, sharkskin was used for smoothing timber in place of sandpaper. Sharks' teeth are specialized denticles.

▶ *The tip of each denticle is made of dentine overlayed with enamel-like vitrodentine. The form of denticles varies depending on where they are found on the shark's body.*

Skeletal system

COMPARE the disposable, razor-sharp teeth of the hammerhead with the deep-rooted, bone-cracking teeth of a *HYENA*. The hyena's teeth have to last it for life, and if they break the animal starves to death.

Most familiar species of fish such as goldfish, salmon, and cod have a skeleton made of bone, which contains a hard reinforcing mineral called calcium phosphate. Fish bone has a lower mineral content than the bones of land animals, so it is not as hard, but it is still bone, nonetheless. By contrast, the skeleton of a shark is made of cartilage, a much softer, more elastic, semitransparent substance. Cartilage is the material that supports the flexible structure of the human nose and ears. It might seem inadequate for supporting the body of a powerful hunter like a hammerhead shark, yet cartilage provides an excellent skeletal material for these fish. In contrast to a land animal, a shark does not need a strong framework to carry its weight. That is because it is supported by the water around it. The main functions of a shark's skeleton are to protect delicate organs like the brain, to stiffen the animal's fins and jaws, and to give its muscles anchorage.

A flexible spine

A shark has a long spine of separate vertebrae, strengthened with a little calcium phosphate to protect the spinal cord. The spine is flexible, so if the shark contracts the muscles on one side, its spine curves in that direction. This also moves its tail fin, which is formed by a vane of cartilage struts and horny rods that attach to an extension of the spine.

The two dorsal fins and anal fin are attached to the spine in the same way as the tail fin. The paired pectoral and pelvic fins are attached to the pectoral and pelvic girdles rather than to

▶ **CARTILAGE**

A shark's skeleton is made of a translucent material called cartilage, which is lighter and less brittle than bone but does not have bone's capacity for regeneration and self-repair.

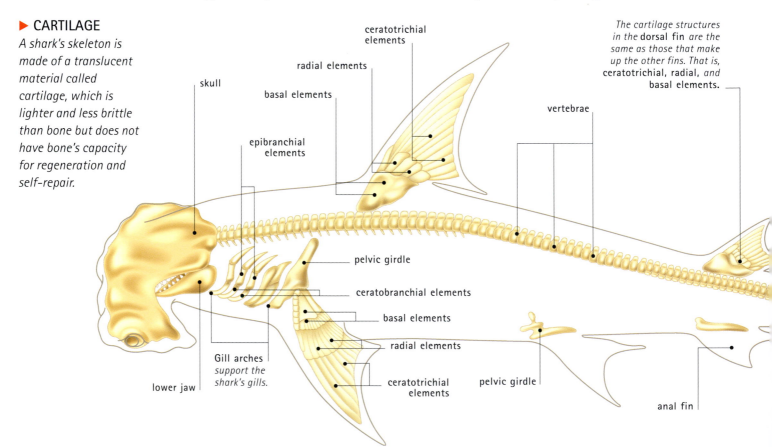

*The cartilage structures in the **dorsal fin** are the same as those that make up the other fins. That is, ceratotrichial, radial, and basal elements.*

ceratotrichial elements

radial elements

basal elements

skull

epibranchial elements

vertebrae

pelvic girdle

ceratobranchial elements

basal elements

radial elements

Gill arches *support the shark's gills.*

lower jaw

ceratotrichial elements

pelvic girdle

anal fin

the spine. The girdles are embedded in the body muscle. Just in front of the pectoral girdle are the gill arches that carry the shark's gills. There are four gills on each side.

Skull and jaws

A hammerhead's skull is made of cartilage, like the rest of its skeleton. With its winglike extensions it is an unusual-looking animal. Apart from these wings, however, and various knobs and struts that act as attachments for the jaw muscles, the skull is a fairly simple brain-containing box. The jaws are separate from this box structure. They are suspended from elastic ligaments at the front, and mobile struts at the back. So when the shark attacks its prey, it can thrust the whole jaw structure forward to seize its victim.

Many sharks have long, pointed teeth ideal for catching their slippery fish prey. In contrast, hammerhead sharks, and other big killers like the tiger shark, have flat triangular or hooked teeth with sharp serrated edges for sawing through skin and flesh. They allow the shark to slice up prey that are too large to swallow whole. The teeth often snap away, but that is not a problem. They are soon replaced by new teeth that move up from inside the shark's mouth. The shark has several rows of teeth in use at any time; since its teeth are constantly being replaced, a shark can get through thousands in its lifetime.

▼ TEETH

Shark teeth undergo a lot of wear and tear. When a tooth breaks, it is replaced by a tooth growing immediately behind it. There is a never-ending supply of replacement teeth growing in rows behind the functioning teeth.

ceratotrichial elements

Muscular system

CONNECTIONS

COMPARE the hammerhead shark's muscular system with that of a *GULPER EEL*. A gulper eel does not have a well-developed muscular system. It lures prey toward itself with its bioluminescent organ and so does not need to move around much.

Hammerhead sharks are active hunters that spend most of their time cruising shallow waters in search of prey. They are constantly on the prowl, so they need a muscle system that enables them to keep moving without getting tired. When they detect a meal they need to launch a swift attack, so their muscles must also be able to deliver a burst of speed when needed. These demands are met by two muscular systems, the red and white muscles.

The red muscle of a hammerhead shark, and many "typical" sharks such as the gray reef shark, lies in a thin layer just beneath its skin. It is red because it is full of blood capillaries and contains a red, oxygen-carrying pigment called myoglobin. The oxygen is vital; the muscle tissue uses it to release energy from sugars

▶ ▼ *The hammerhead's muscular system is able to provide power for slow movement over very long distances as well as short-distance rapid movement for attacking fleeing prey.*

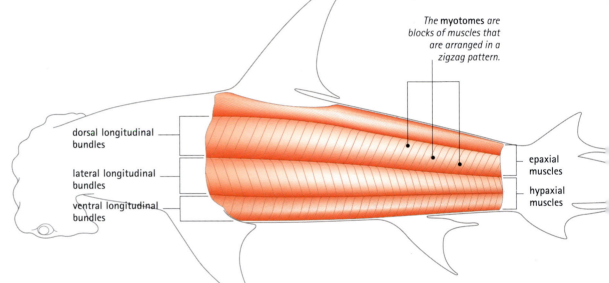

The **myotomes** *are blocks of muscles that are arranged in a zigzag pattern.*

dorsal longitudinal bundles

lateral longitudinal bundles

ventral longitudinal bundles

epaxial muscles

hypaxial muscles

SEGMENT OF SHARK MUSCLE SHOWING FIBER TYPES

CROSS SECTION THROUGH SHARK'S TORSO

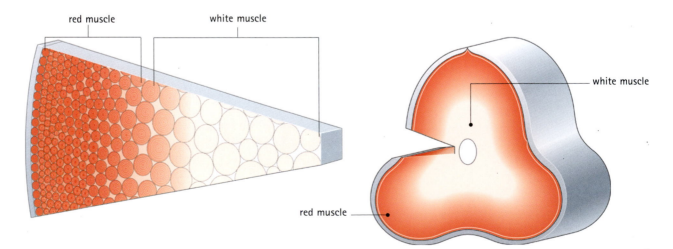

red muscle

white muscle

white muscle

red muscle

Streamlined swimmers

Most active, predatory sharks swim in the same way as hammerheads. However, some mackerel sharks, such as the great white shark, are different. The great white's body is more streamlined, being pointed at both ends; it holds its body stiffly and propels itself with rapid, powerful strokes of its tail fin. It has much more red muscle than a typical shark; this lies inside the white muscle, near the spine. This arrangement allows the great white to cruise at higher speeds than most other sharks, and to roam over vast distances in search of prey.

▶ *Great white sharks are highly nomadic and travel long distances in search of prey such as seals, turtles, and belugas.*

provided by the blood. Red muscle can keep working steadily for hour after hour, day after day. It has virtually limitless stamina.

White muscle

The much thicker muscle layer lying inside the red muscle is formed of white muscle. It is white, or at least pale, because it has no myoglobin and a poor blood supply. It is powered by a starch called glycogen, which is formed by linking glucose molecules into chains, and acts as an energy store in muscles. Energy production in the white muscle does not require oxygen. This system produces a lot of energy in a very short time, making white muscle ideal for use during the rapid pursuit of fleeing prey. However, using white muscle leads to the formation of a by-product called lactic acid. This acid soon builds up the muscular system, causing exhaustion. Consequently, the red muscle is used for endurance and the white muscle for sprinting.

Power packs

The body muscles are arranged in vertical, zig-zag blocks all the way down the shark's flanks. As the shark swims, its muscles contract in sequence, producing a wave effect that passes back along the shark's body to its tail. The wave acts against the resistance of the water to drive the shark along. Most of the time the shark moves with a steady, graceful action, powered by the tireless red muscle, but when the shark launches an attack the white muscle blocks surge into action, allowing the animal to move at up to 20 mph (32 km/h).

The pelvic fins and long pectorals are controlled by muscles that can alter their angle, rather like the control surfaces of an airplane. This enables the shark to maneuver up and down in the water, and to make tight turns as it chases prey.

Jaw power

Hammerhead sharks often feed on large prey that cannot be swallowed whole, so they need enough muscle power in their jaws to slice it up. Biologists have shown that a big shark like the great hammerhead can bite with a force of about 44,000 pounds per square inch (3,200 kg/cm^2). By comparison, human jaws can muster only 150 pounds per square inch (10 kg/cm^2). The shark's bite translates to a massive 132 pounds (60 kg) of force for each of its teeth. This explains why sharks lose so many teeth to fracture or breakage.

Nervous system

COMPARE the well-developed lateral line system of the hammerhead shark with the system of the *GULPER EEL*. The gulper eel relies heavily on sensitivity to vibration to detect prey in the total darkness of the deep ocean.

Sharks are notorious for their killing efficiency, but their real talent is for finding prey in the first place. Their nervous systems are tuned to detect the slightest hint of a possible meal, and to locate it with pinpoint accuracy even in dark, murky waters.

Super smellers

A shark's most important sense over distance is its sense of smell. Sharks have an amazing ability to track scents that interest them, being able to detect one part of blood in 10 billion parts of water. That is equivalent to roughly one drop of blood in a large swimming pool.

A shark detects scent using its nostrils. Unlike the nostrils of mammals, shark nostrils do not lead to respiratory organs but are blind sacs lined with microscopically folded sensory tissue. Each nostril is shaped so that water flows in one side and out the other. The sensory tissue samples the water and detects any chemical content, converting the information into electrical nerve signals. These pass to the brain, where the scent is matched to a mental database of edible substances. If it triggers a response, the shark follows the scent trail to its source. Sharks have been known to track down prey in this way from a distance of more than half a mile.

The nostrils of a hammerhead shark are at each end of the cephalofoil. The wide spacing allows minute differences in the scent concentration at each nostril to be detected. This helps the animal home in on its target— the spacing of your ears enables you to locate the source of a sound in much the same way.

Pressure sensors

A shark can detect sounds, too, both through its ears and through the pressure-sensors of its lateral line system. Its ears function mainly as orientation sensors, so the shark knows which

► *The hammerhead shark finds prey using its nervous system and sensory organs. Over long distances the shark is most dependent on its sense of smell, but up close the shark finds prey using its sensitivity to electric fields.*

IN FOCUS

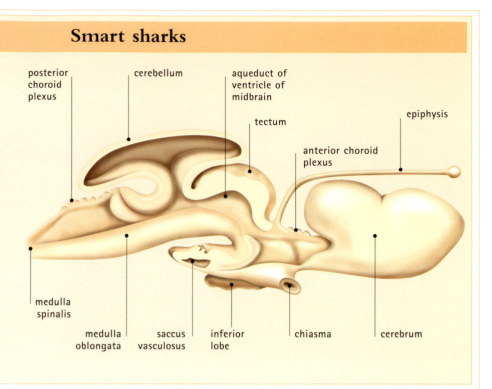

Smart sharks

Sharks are often described as instinctive killing machines, but that underrates their intelligence. Sharks can learn by experience, and they devise hunting strategies for outwitting intelligent animals such as sea lions. Some sharks, including the great white shark, may even hunt cooperatively, but scientists are not agreed on that theory. Great white sharks sometimes travel together to hunting areas but probably hunt separately when they get there.

▶ **SECTION OF BRAIN**

Shark brains are large compared to other fish and have an elongated form. From above they appear Y-shaped.

posterior choroid plexus — cerebellum — aqueduct of ventricle of midbrain — tectum — anterior choroid plexus — epiphysis

medulla spinalis — medulla oblongata — saccus vasculosus — inferior lobe — chiasma — cerebrum

way up it is. However, they are also sensitive to low-pitched sounds, which travel faster through water than on land, and over distances of several miles. They are often the first thing to alert a shark to a possible meal.

The lateral-line system extends along a shark's flanks and over its head. It is a feature of many types of fish, but it is particularly well developed in hammerhead sharks. It consists of a series of fluid-filled canals just below the skin connected to the surrounding water by tiny pores. The canals contain sensory cells with tiny hairlike projections. When the "hairs" are disturbed by changes in water pressure they send nerve impulses to the brain.

The system can detect pressure changes created by other animals passing by in the dark, or by the shark's swimming close to an obstacle. It can also detect low-frequency vibrations produced by the struggles of injured fish.

Shark vision

The eyes of a shark are more efficient than those of many fish. They have a mirrorlike layer at the back that reflects light into the cells of the retina, like the eyes of a cat. This improves their eyes' sensitivity at the low light levels that

▼ **OLFACTORY ORGAN**

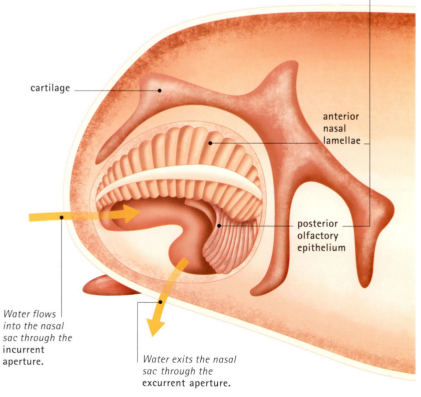

Chemicals in seawater are detected as odors by cells in the anterior nasal lamellae and the posterior olfactory epithelium, which are covered with millions of olfactory cells.

cartilage — anterior nasal lamellae — posterior olfactory epithelium

Water flows into the nasal sac through the incurrent aperture.

Water exits the nasal sac through the excurrent aperture.

COMPARATIVE ANATOMY

Chimeras

The chimeras are cartilaginous fish like sharks, and share many of their features. They are adapted for life in deep, cold waters, and several species have a long snout that is richly endowed with chemical and electrical receptors (ampullae of Lorenzini) for locating prey. Some have a strange, spooky appearance, partly due to their very large, iridescent turquoise eyes. These are much bigger than those of typical sharks. The giant eyes are used to gather as much light as possible in the ocean depths, where only the faintest glimmer of light filters down from the surface.

◄ *The elephant fish is a chimera whose common name comes from its trunklike snout, which is covered with electroreceptors.*

exist underwater. However, even in the perfectly clear waters of tropical oceans a shark cannot see much farther than about 100 feet (30 m), so when hunting it relies on its other senses until it gets close to its target.

Looking sideways

The eyes of a hammerhead shark are widely spaced at the tips of the cephalofoil, so it might be expected that these animals would have outstanding stereoscopic vision (the ability to see in depth and judge distances accurately). However, each eye faces sideways, and the hammerhead cannot swivel them forward to face in the same direction. It literally cannot see what is in front of its nose, and has no stereoscopic vision whatsoever.

If this were a problem for hammerheads, the position and structure of the eyes might have evolved in a completely different way. Yet although head shape in hammerheads varies a lot, from the narrow bonnethead to the broad winghead shark, their eyes always face sideways. That is because hammerheads and other sharks do not rely on vision when they close in on prey. They depend on another sense beyond the scope of human experience—electrodetection.

Betrayed by nerves

All sharks and rays are extremely sensitive to electrical impulses. The skin of a shark's head has gel-filled pores containing electroreceptors called the ampullae of Lorenzini. These extensions of the lateral-line system are able to detect an electrical field of less than five billionths of a volt. The ampullae are by far the most sensitive electrical detectors in the animal kingdom. Provided the shark is close enough, it can sense two types of electric fields. One is generated by the difference of ion concentration between the body of the prey and the seawater surrounding it. The other field is produced by the activity of the muscular system within the animal.

The cephalofoil of a hammerhead shark is peppered with electroreceptors. As a shark swims toward its intended prey, the width of its head ensures that some of the cells pick up a stronger signal than others, so it can turn in the direction that equalizes the signals, just as it does with scents picked up by its widely spaced nostrils. The system is especially useful when the shark is hunting animals that are buried in the sand of the seabed; the shark uses its electroscanning cephalofoil like a mine detector to find them.

IN FOCUS

Why the weird shape?

Researchers have puzzled for years over the function of the hammerhead's cephalofoil. Suggestions that it serves a hydrodynamic function and allows the animal to make tight turns in the water have been proved incorrect. The cephalofoil's primary role seems to be sensory. It bears vast numbers of ampullae of Lorenzini in just the place they are most needed. This arrangement gives the hammerhead an electrical detection system that is vastly superior to that of other sharks. Hammerheads can detect faint electrical impulses from 50 percent farther away than the maximum distance that other sharks can manage. The cephalofoil also allows the sharks to derive detailed information on the location of an electric field source. Stability conferred by the cephalofoil's shape may also help the head remain level as the animal turns, allowing it to maintain a clear electrical picture of the object of interest.

▶ **AMPULLAE OF LORENZINI**
Hammerhead shark
The ampullae of Lorenzini detect the electric fields produced by an animal's nervous system. Sharks use this information to track down prey.

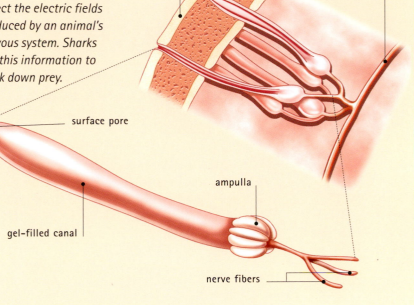

epidermis

nerve fiber

surface pore

gel-filled canal

ampulla

nerve fibers

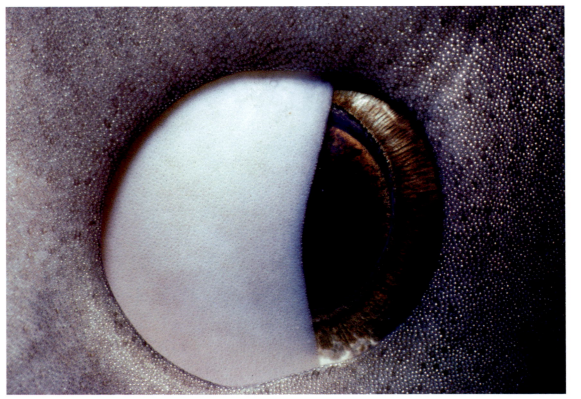

◀ *The nictating membrane of a tiger shark's eye helps keep the eye from being damaged when the shark attacks prey.*

471

Circulatory and respiratory systems

CONNECTIONS

COMPARE the hammerhead shark's respiratory system with that of a *HAGFISH*. Water enters the hagfish through its nostrils rather than through the mouth. The hagfish is also capable of breathing through its skin.

▼ *Sharks have four-chamber hearts, but they are very different to the four heart chambers in a mammal. The heart pumps blood containing nutrients and oxygen around the body to the organs and tissues.*

As a hammerhead shark cruises through the ocean, its muscles use energy. Energy is usually produced by a process called aerobic respiration. Glucose sugar in a cell reacts with oxygen. This yields carbon dioxide gas, water, and energy. So the shark needs a steady supply of oxygen, and some way of getting rid of the waste carbon dioxide.

Ocean water contains plenty of dissolved oxygen, especially near the surface. The shark gathers it with its gills, clusters of feathery blood-filled tubes with extremely thin walls that lie in the head. There are four rows of gills on each side, attached to strong gill arches of cartilage at the back of the shark's mouth. As the shark swims forward, water flows in through its mouth, over the gills, and out through five gill slits on each side. Meanwhile,

COMPARATIVE ANATOMY

Warm-blooded sharks

Fast-swimming mackerel sharks such as the great white shark are effectively warm-blooded. Heat generated by the active red muscle near the shark's spine warms the blood passing through it. As the warm, deoxygenated blood returns through veins to the shark's heart, its heat passes into nearby arteries containing oxygenated blood being pumped to the muscle. This heat-exchanging arrangement ensures that the muscle is kept at a higher temperature than the surrounding water, improving its efficiency.

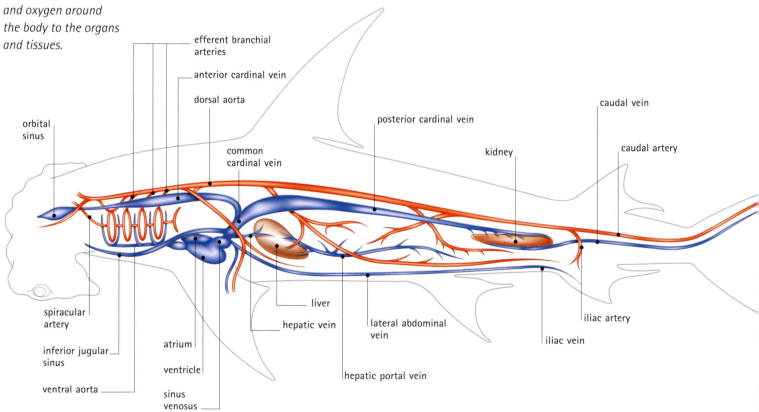

efferent branchial arteries

anterior cardinal vein

dorsal aorta

posterior cardinal vein

caudal vein

orbital sinus

common cardinal vein

kidney

caudal artery

spiracular artery

liver

lateral abdominal vein

iliac artery

inferior jugular sinus

atrium

hepatic vein

iliac vein

ventral aorta

ventricle

hepatic portal vein

sinus venosus

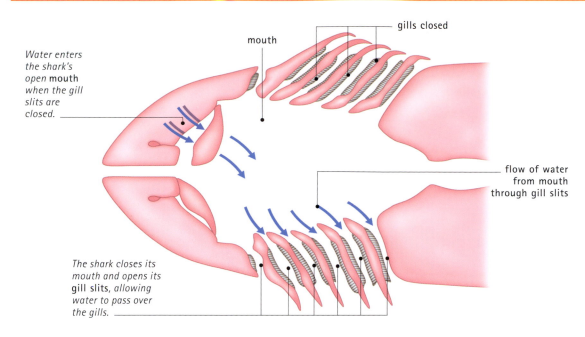

mouth

gills closed

Water enters
the shark's
open mouth
when the gill
slits are
closed.

flow of water
from mouth
through gill slits

The shark closes its
mouth and opens its
gill slits, allowing
water to pass over
the gills.

◀ GILLS

*A shark obtains oxygen
from water using
organs called gills. With
the gills closed, water
enters the shark's open
mouth. The shark then
closes its mouth and
opens the gills, allowing
the water to pass
across the gills and
out of the body via
the gill slits.*

blood carrying waste carbon dioxide from the shark's body is pumped through the gills. The carbon dioxide passes out through the thin walls of the gills and into the water, while oxygen passes in from the water. The oxygenated blood then flows to the shark's muscles and organs.

The faster the shark swims, the more oxygen it uses. However, since the water also flows through its gills faster, this automatically supplies the extra oxygen it needs. Comblike structures on the gill arches, called gill rakers, trap any debris in the water to prevent it from damaging the delicate gill filaments. If the shark stops swimming, it can pump the floor of its mouth to push water through its gills and prevent itself from suffocating.

Blood is pumped to the shark's gills by its four-chamber heart. After passing through the gills, the oxygenated blood flows through arteries directly to the organs and tissues where it is needed. The oxygen is used and replaced in the blood by waste carbon dioxide. The blood returns through veins to the heart, which pumps it back to the gills.

A shark's heart is small compared with its body size, but its pumping power is supported by the squeezing effect of the animal's muscles as it swims. Many sharks must therefore keep moving to maintain a good flow of blood through the body. Blood flow is enhanced by movement in many animals. For example, it explains why soldiers on parade occasionally faint after standing still for a long time.

IN FOCUS

Angel shark spiracles

Angel sharks and other bottom-living sharks, which spend much of their lives half-buried in the sand of the seabed, cannot use their mouth to draw oxygen-rich water over their gills. Instead, these species have openings behind their eyes called spiracles, which draw water in from above. Sharks with spiracles have an enhanced blood circulation to the eyes and brain. This helps ensure that these sensitive organs never run short of oxygen.

◀ *One spiracle is
visible on this angel
shark. The spiracle is
the white crescent-
shaped mark behind
this shark's eye.*

Digestive and excretory systems

Like all big predators, hammerhead sharks usually eat large meals but with long intervals between them. Prey is often scarce, and when a shark finds food it must eat as much as possible to see it through to the next meal. It can do this because it has an elastic, expandable stomach that can hold up to 10 percent of the shark's body weight—roughly five times its daily food requirement.

Food in the stomach is broken down by digestive juices, and the products pass into the intestine to be absorbed. The processed food passes slowly over the lining of the intestine, guided by a corkscrew-shaped spiral valve, which makes the food travel around the intestine rather than straight through it. This ensures that the nutrients have plenty of time to be absorbed into the shark's bloodstream.

Buoyancy aids

A lot of the nutrients taken up by the shark's blood are stored in its large liver as glycogen. This can be converted to glucose for use by cells later. The liver also destroys toxins in the

blood, and turns some waste products into substances that can be filtered out of the shark's blood when it passes through the kidneys.

The liver also breaks down fats absorbed at the intestine and converts them into oil. The oil is lighter than water; the liver contains so much oil that it offsets the shark's weight, giving the shark the near-neutral buoyancy that it needs to move easily through the ocean.

PREDATOR AND PREY

Giant plankton eaters

Like most big sharks, a hammerhead is a predator at the top of the food chain and eats large fish, squid, and similar animals. The biggest sharks of all have a very different diet. The massive basking shark and whale shark feed on huge quantities of krill and small shoaling fish. They cruise through the shoals with their mouth gaping wide open, trapping the animals in their sievelike gill rakers. This system is so efficient that the whale shark grows larger than any other fish, sometimes reaching more than 40 feet (12 m) long.

▼ SPIRAL VALVE

The helical shape of the valve in a shark's hind-gut lengthens the path of digestion. The shape maximizes the absorption of nutrients from digested food.

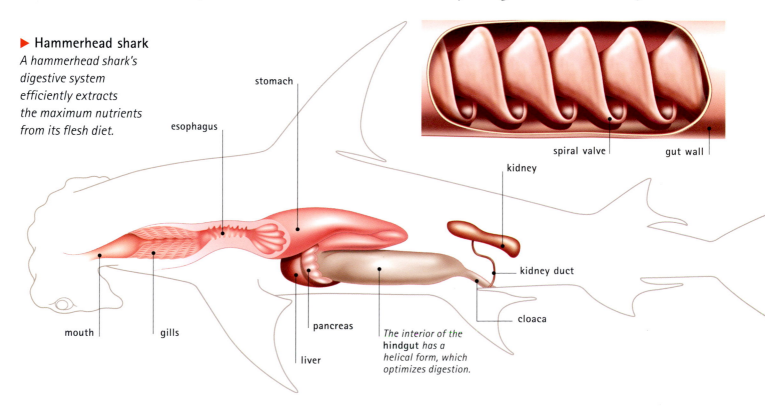

► Hammerhead shark

A hammerhead shark's digestive system efficiently extracts the maximum nutrients from its flesh diet.

esophagus

stomach

mouth

gills

pancreas

liver

The interior of the **hindgut** *has a helical form, which optimizes digestion.*

kidney

spiral valve

gut wall

kidney duct

cloaca

Reproductive system

Fertilization in all sharks is internal. The male clings to the female—often using his teeth—and introduces sperm into her body using one of the two long "claspers," which lie close to his pelvic fins. The sperm pass down a groove in the clasper into the female's two-part uterus. There it fertilizes the eggs. Internal fertilization allows the female to produce much bigger, more nutrient-rich eggs. This in turn increases the chances of survival of the young.

Some species, such as bullhead sharks, lay their fertilized eggs in protective egg cases often known as mermaid's purses. Each contains one egg that develops into a baby shark, or pup, inside the case and then hatches. The eggs of other sharks, such as mackerel sharks, open inside the female's body. The pups stay in her uterus, nourished by the yolk of the egg, until they are ready to be born.

Hammerhead sharks give birth to up to 40 pups. The developing young are supplied with nutrients from the mother's bloodstream. The

◀ *Male sharks can be easily identified by their two backward-pointing claspers, which lie close to the pelvic fins.*

nutrients are delivered to each unborn pup through a placenta and umbilical cord similar to that of a placental mammal.

TREVOR DAY

FURTHER READING AND RESEARCH

Ashley, L. M. 1988. *Laboratory Anatomy of the Shark.* McGraw-Hill: Columbus, OH.

Mallory, K. 2001. *Swimming With Hammerhead Sharks.* Houghton Mifflin: Boston, MA.

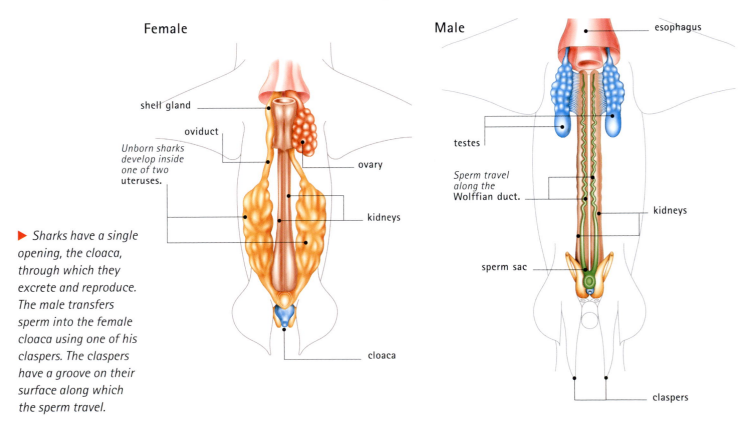

Female

- shell gland
- oviduct
- *Unborn sharks develop inside one of two uteruses.*
- ovary
- kidneys
- cloaca

▶ *Sharks have a single opening, the cloaca, through which they excrete and reproduce. The male transfers sperm into the female cloaca using one of his claspers. The claspers have a groove on their surface along which the sperm travel.*

Male

- esophagus
- testes
- *Sperm travel along the Wolffian duct.*
- kidneys
- sperm sac
- claspers

Hare

ORDER: Lagomorpha FAMILY: Leporidae GENUS: *Lepus*

Hares and their relatives live in a wide variety of habitats in many regions of the world. They inhabit bush and deserts, forests and swamps, grasslands and fields, and tundra and mountains. They live from Alaska to Argentina, from Japan to western Europe, and in Africa. Hares were introduced to Australia by people. Hares and rabbits, together with their smaller relatives pikas, form the order Lagomorpha and are sometimes called lagomorphs.

Anatomy and taxonomy
Scientists group all organisms into taxonomic groups based largely on anatomical features. Together with rabbits, hares belong to the family Leporidae. Their closest relatives, the pikas, make up the family Ochotonidae.

● **Animals** Animals are multicellular organisms that usually get the nutrition they need by consuming other organisms, including plants and other animals. Animals differ from other multicellular life-forms in their ability to move from one place to another, usually by using muscles. Most animals have a nervous system that allows them to react rapidly to touch, light, and other stimuli.

● **Chordates** At some time in its life cycle a chordate has a stiff, dorsal (back) supporting rod called the notochord, which runs along most of the length of the body.

● **Vertebrates** The vertebrate notochord develops into a backbone (spine, or vertebral column) made up of units called vertebrae. The muscular system that moves the head, trunk, and limbs consists primarily of muscles in a mirror-image arrangement on either side of the backbone. This arrangement is called bilateral symmetry.

● **Mammals** These are endothermic (warm-blooded) vertebrates with hair or fur made of keratin. Females have mammary glands that produce milk to feed their young.

▶ *Several lagomorph species are so different anatomically from others that they are placed in genera of their own. Although they are a diverse group, little is known of the relationships or biology of many lagomorphs. The Sumatran short-eared rabbit, a beautiful striped species, has been observed in its forest habitat just a few times since the early 20th century. A similar species, the Annam striped rabbit of Vietnam, was discovered as recently as 1999.*

Animals
KINGDOM Animalia

Vertebrates
SUBPHYLUM Vertebrata

Mammals
CLASS Mammalia

Glires
SUPERORDER Glires

Rodents
ORDER Rodentia

Lagomorphs
ORDER Lagomorpha

Pikas
FAMILY Ochotonidae

Hares and rabbits
FAMILY Leporidae

Pikas
GENUS *Ochotona*

Short-eared rabbits
GENUS *Nesolagus*

Ryuku rabbit
GENUS AND SPECIES
Pentalagus furnessi

Bristly rabbit
GENUS AND SPECIES
Caprolagus hispidus

Red rabbits
GENUS *Pronolagus*

Central African rabbit
GENUS AND SPECIES
Poelagus marjorita

Bushman hare
GENUS AND SPECIES
Bunolagus monticularis

European rabbit
GENUS AND SPECIES
Oryctolagus cuniculus

Volcano rabbit
GENUS AND SPECIES
Romerolagus diazi

Cottontails
GENUS *Sylvilagus*

Pygmy rabbit
GENUS AND SPECIES
Brachylagus idahoensis

Hares and jackrabbits
GENUS *Lepus*

In mammals, the lower jaw is formed by a single bone, the dentary, which hinges directly to the skull. A mammal's inner ear contains three small bones. Mammalian red blood cells, when mature, lack a nucleus; all other vertebrates have red blood cells that contain a nucleus.

● **Placental mammals** Placental mammals, or eutherians, nourish their unborn young through a placenta, a temporary organ that forms in the mother's uterus, or womb, during pregnancy.

● **Glires** Glires (rodents and lagomorphs) are gnawing animals with teeth that grow constantly. The front teeth (incisors) are long and can project out of the mouth even when the animal has its mouth closed. Glires lack canine teeth, and in the region of the jaw where canines might grow in other animals they have a gap called the diastema.

● **Rodents** Rodents, such as mice and rats, have jaws of equal width that line up on both sides of the mouth at once. Rodents typically have relatively long back legs and many have prominent ears, but none of them has as extreme a body profile as that of a lagomorph.

● **Lagomorphs** There are 47 species of hares, rabbits, and pikas in 11 genera (groups of related species). Compared with rodents, lagomorphs have an extra pair of peglike incisor teeth on the upper jaw. They have tooth enamel covering the whole of their long incisors, not just the front face of each tooth, as with rodents. The upper jaw of a lagomorph is wider than the lower, so it can chew on only one side of the mouth at a time.

● **Pikas** Pikas have short, round ears and are smaller than rabbits and hares. They utter piercing whistles when they sense danger. Unlike their larger relatives, pikas have four limbs of similar length. They do not run fast and never stray far from a hideaway.

▲ *Its very long ears and, when it is running, its long hind legs are a brown hare's most obvious external body features.*

● **Leporids** Hares and rabbits both have long ears, powerful back legs, long incisor teeth, and a split lip. Although they are closely related, there are differences in anatomy and behavior between rabbits and hares. Rabbits are typically smaller than hares and have smaller hind limbs. They dig burrows for shelter and give birth underground to naked, blind, helpless babies. Hares have longer ears and live above ground. They escape danger by running, powered by very long back legs. Baby hares are born above ground. They are furred and ready to run within minutes of birth. The brown hare lives in Europe and is one of around 23 species of hares in the genus *Lepus* (hares and jackrabbits).

FEATURED SYSTEMS

EXTERNAL ANATOMY The brown hare has prominent ears. Long, heavily muscled back legs contrast with much shorter front limbs. The tail is short. *See pages 478–481.*

SKELETAL SYSTEM Hares' teeth grow constantly from the roots, and there are long incisors at the front of the jaw. The hind limbs are very long. *See pages 482–483.*

MUSCULAR SYSTEM Powerful muscles drive the back legs, providing the thrust that makes the brown hare one of the fastest-running animals on Earth. *See pages 484–485.*

NERVOUS SYSTEM The small brain weighs just 0.35 to 0.45 ounce (10–13 g). *See page 486.*

CIRCULATORY AND RESPIRATORY SYSTEMS Oxygen passes through the lungs to the blood, which the heart pumps around the body. *See pages 487–488.*

DIGESTIVE AND EXCRETORY SYSTEMS Hares and their relatives are hindgut fermenters: they have bacteria in their hindgut that digest their plant food. After one pass through the gut, hares consume their food for a second time and so digest it twice. *See pages 489–491.*

REPRODUCTIVE SYSTEM Females have a double uterus, enabling them to become pregnant again before they give birth to the previous set of baby hares. *See pages 492–493.*

External anatomy

CONNECTIONS

COMPARE the large ears of a hare with those of a mammal such as a *DOLPHIN*, an *ELEPHANT*, or an *OTTER*.

COMPARE the hare's legs, especially its long hind legs, with those of a rodent, such as a *RAT*.

There are no other animals quite like rabbits and hares. From their tall upright ears to their large eyes, and from their convex back profile to their outsized back legs and tiny tail, lagomorphs have a unique shape. Indeed, the name *lagomorph* means "hare-shaped."

Lagomorphs are medium-sized animals. Pikas are smallest, 4.7 to 11.2 inches (12–28.5 cm) from nose to hind toe. The brown hare is the largest at 20 to 30 inches (50–76 cm). Among lagomorphs, unlike most other mammals, females are larger than males.

The brown hare's huge ears, large eyes, and sensitive nose are its early warning systems and vital to its survival. It keeps a low profile when feeding. The eyes are high on the head, with one eye pointing right and the other left, giving almost all-around vision. Scent signals

▼ **Brown hare**
All species of true hares share the main features of their anatomy. There is, however, a marked variation in ear length, and there are some differences in fur coloration. The brown hare has some of the longest ears of any lagomorph.

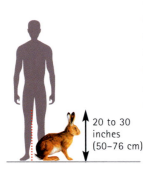

20 to 30 inches (50–76 cm)

*Even when the hare is feeding, its long **ears** can detect distant sounds. The ears also help the hare cool itself in hot weather.*

*The **fur** is soft and dense. The underfur traps air close to the skin and so keeps the animal from getting too cold in winter.*

*The large bulging **eye** on the side of the head gives a wide field of view.*

*The **nose** has slitlike nostrils that can be opened and closed by a fold of skin above.*

*Each **hind leg** is very long with a long foot and five toes.*

*A hare can rotate its head to the left and right on its **neck**.*

foreleg

*There are only four visible **toes** on each foreleg. The fifth toe is tiny.*

*The short white **tail** is held down when the hare is running.*

478

are important to lagomorphs, and they twitch their noses constantly, smelling the air for danger. Hares can close their slitlike nostrils.

The whiskers of hares, rabbits, and pikas have a sensory function. The whiskers spread as wide as the animal's body and are sensitive to touch. Like other whiskered animals, hares can tell in advance if they will fit through a gap by feeling how much their whiskers touch their surroundings.

Directional ears

The brown hare's elongated ears gather sounds in the same way that satellite dishes gather television signals. The ears' position high on the head and their large size, directional movement, and funnel shape combine to enable the hare to hear all around when feeding with its head down. Pikas can close their ear openings to keep out wind and rain in bad weather. Hares may also lay their ears flat against their back when resting or hiding.

Large ears also enable lagomorphs to lose heat from their body. The lagomorph with the largest ears of all is the antelope jackrabbit. It lives in the southern deserts of the United States. Its colossal ears have thin skin covering lots of blood vessels. Heat escapes from the blood through the skin, keeping the animal from overheating. Hares living in cooler environments, such as the arctic hare, have smaller ears that do not lose so much heat. A brown hare's ears help it cool down after exercise. When the hare comes to rest, its heart is still pumping blood quickly. The extra heat generated by running is lost through the ears.

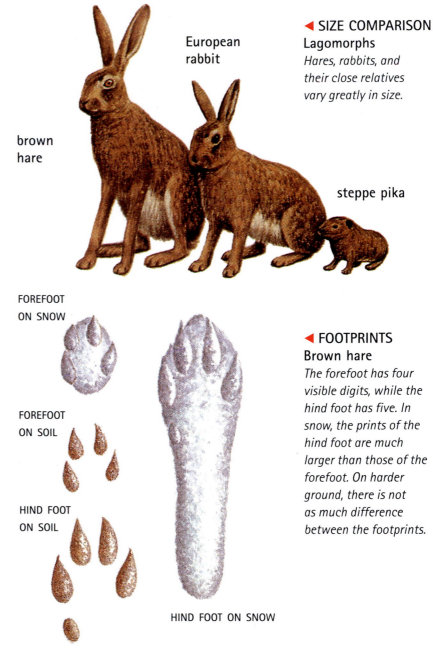

◄ SIZE COMPARISON
Lagomorphs
Hares, rabbits, and their close relatives vary greatly in size.

European rabbit

brown hare

steppe pika

FOREFOOT ON SNOW

FOREFOOT ON SOIL

HIND FOOT ON SOIL

HIND FOOT ON SNOW

◄ FOOTPRINTS
Brown hare
The forefoot has four visible digits, while the hind foot has five. In snow, the prints of the hind foot are much larger than those of the forefoot. On harder ground, there is not as much difference between the footprints.

antelope jackrabbit

black-tailed jackrabbit

snowshoe hare

Arctic hare

◄ EAR SHAPE
Hares
All hares have long ears, but those living in hot regions usually have the longest. The ears of the antelope jackrabbit, which lives in the Sonoran Desert, Arizona, are up to 8 inches (20 cm) long. They help keep the animal cool in the desert heat.

479

COMPARATIVE ANATOMY

Pikas

Pikas are rabbits' and hares' closest relatives. There are 29 species, all in one genus, *Ochotona*. Pikas live in rocky areas in mountains in western North America and northern Asia, where their range also extends down to sea level. Pikas also once lived in Europe. Pikas' legs are more equal in size than those of rabbits and hares,' and they have smaller, rounder, more mouselike ears than other lagomorphs. When pikas spot danger, they dive quickly for cover between rocks or run down their burrow when they spot danger. Pikas collect and dry food to store for the winter. Like other lagomorphs, they project their incisors through the split upper lip, enabling them to carry their food without keeping their mouth open. Pikas are the most vocal lagomorphs, giving a high-pitched whistle when they spot danger.

A split lip

Human lips can seal the mouth. Lagomorphs have lips that do not seal. They have a split upper lip that parts in the middle of the front of the mouth. It looks like a pair of curtains open at the bottom and closed at the top. The gap in the lips reveals the incisors inside and enables lagomorphs to gnaw food and carry items such as food and bedding without opening their mouth fully. The items fit behind the incisors. The lips fit behind the objects and hold them in place.

Rabbits and hares have more movement in the neck than rodents. Behind the neck, the slightly arched back ends in a short tail, or scut. Lagomorphs have a fluffy white scut. They use it to make visual signals to other individuals of their species. A flashing white tail at the back of a running rabbit may also confuse predators as the lagomorph runs away at speed. Hares, however, point their tail downward when running, concealing the white fur under the tail and revealing the black upper surface.

Legs and feet

Rabbits and hares have front and back legs of different lengths. The forelimbs are short and straight, ending in small furred and padded feet with a stout claw at the end of each of the four visible toes (the fifth toe is tiny). The hind legs are far larger and each has five toes on a much longer foot. The long hind feet provide more traction (grip) and a larger surface area against which to push. The forelegs provide balance

▶ RUNNING ACTION
Brown hare
When a hare hops or runs, its long hind legs move in front of the forelegs, allowing the animal to make a powerful leap. This action is more pronounced when the hare runs fast, giving it even more propulsion.

HOPPING

RUNNING

and take the weight of the animal while the back legs swing forward into position for the next push. Hares and rabbits place their forefeet on the ground one at a time, leaving a distinctive footfall of two hind feet next to each other and two forefeet placed one behind the other.

Warmth-retaining fur

The brown hare has soft, dense fur. The undercoat traps air next to the skin and keeps the animal warm. Longer guard hairs lie on top. They add strength and waterproofing to the pelage (coat). Hares need effective weatherproofing to survive in open places. Rabbits have softer fur than hares but spend less time outside exposed to the weather. All rabbits and hares have hairy soles on their feet. The hairs help with grip and provide insulation for those animals living in extreme environments. Hairy soles offer protection from the extreme heat of desert sand and the cold of tundra snow.

Color changes

Most lagomorphs are brown, darker on top and paler underneath. Many have black fur at the extremities, such as the upper tail and the tips of the ears. Some have black markings elsewhere. All hares molt their fur seasonally in response to reliable indicators, such as day length. In most species, summer and winter coats are the same color, but not all hares stay brown all year. The mountain hare of Europe, the North American snowshoe hare, and the arctic hare all live in cold regions. These species remain active in winter and change color with the seasons, enabling them to match their surroundings. These hares have gray-brown fur in summer that provides camouflage among plants and rocks. A predator would have an easy time finding dark hares against a white background if they kept their summer coat in the snowy winter.

In the fall, hares that live in cold regions molt into a white winter coat and are able to hide more safely in the snow. In spring they molt again to their dark coat. The color of an arctic hare's winter coat is controlled by the average temperature at the time of the molt. If the weather is cold, the winter coat is white; if warm, the fur will be brown.

CLOSE-UP

Hollow-fiber insulation

Hare hair has a hollow center filled with pigments that color the coat. White fur has no pigments inside the hair and is hollow. The white color is caused by a trick of the light, not pigments. The coat looks white because of the way that the hairs reflect light. Tubular white hairs might also explain why some hares in cold regions turn white in winter. Hollow fibers are more insulating, and so warmer, than solid ones. Also, more sunlight passes through a white coat than a dark one and warms the skin directly.

▼ *In the depths of winter, a snowshoe hare's white fur gives the animal good camouflage in a snowy landscape. In spring the hare will molt its white coat and replace it with gray-brown fur.*

Skeletal system

The skull of a hare has a large, curved nasal bone that protects the sensitive smelling apparatus. The bone opens widely at the front for the large nostrils. Large, curved bone extensions called supraorbital processes project from the skull above each eyeball and protect the eyes. Rabbits have an area of spongy bone lining their upper jaw. This may allow heat to escape from the animal as it runs, keeping it cool. The large orbit (eye socket) is high on the skull, pointing sideways to accommodate the eye. There is a large exit hole at the back of the orbit, through which the optic nerve passes to the brain. Only a thin section of bone in the middle of the skull separates the left eye socket from the right. The brain cavity is small.

Unusual teeth

Lagomorphs have a pair of long incisor teeth that project from each jaw at the front of the mouth. They also have a peglike incisor tooth on each side of the upper pair of jaws. Hares and rabbits differ from rodents in that their teeth are covered on all sides by enamel. Rodents have thick enamel on the front face of their incisor teeth but none on the rear. The rear surface wears more quickly than the front. Between the front and rear is a middle layer, which is softest of all. This layer wears down fastest. This pattern of wear leaves a chisel-like cutting edge at the front of the incisors. Lagomorphs' teeth wear more evenly than those of rodents and are thus never as sharp.

Behind the front teeth is a large gap called the diastema; then there is a row of premolars and molars on each jaw. Hares eat a lot of tough plants, including grasses, which contain hard shards of silica (a glasslike mineral). The

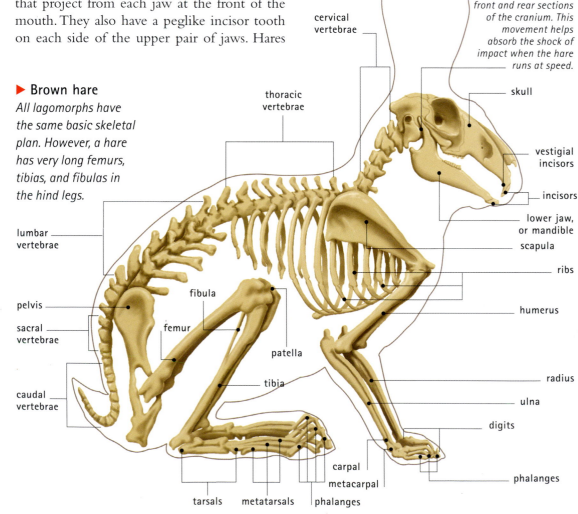

▶ Brown hare
All lagomorphs have the same basic skeletal plan. However, a hare has very long femurs, tibias, and fibulas in the hind legs.

The intracranial joint is a suture that allows movement between the front and rear sections of the cranium. This movement helps absorb the shock of impact when the hare runs at speed.

cervical vertebrae

thoracic vertebrae

lumbar vertebrae

pelvis

sacral vertebrae

fibula

femur

patella

caudal vertebrae

tibia

tarsals

metatarsals

phalanges

carpal

metacarpal

skull

vestigial incisors

incisors

lower jaw, or mandible

scapula

ribs

humerus

radius

ulna

digits

phalanges

shards wear the teeth constantly as the animal eats. Incisors are cutting teeth. Slicing through plants wears a lagomorph's constantly growing incisors at the tips. The peg incisor teeth are probably vestigial features; that is, they have been inherited from ancestral hares but no longer serve any purpose.

A hare can move its jaws up and down and side to side. Molars in the cheek grind the food. When it chews, a hare moves the larger molars on the lower jaw up against the upper molars. Sideways grinding movements crush the food into more digestible pulp. Each molar has a rough upper surface, crisscrossed with ridges that improve grinding ability.

An articulated skull

Bounding at full speed, a hare flexes its spine so its hind feet reach farther forward than the forelegs. The release of energy as the spine straightens and hind legs spring from the ground forces the animal forward. The spine then flexes in the opposite direction. At full speed both the pelvis and the head may rise higher than the middle of the back. The hare needs to see clearly where it is going, so it keeps its head as level and as still as possible. This would be impossible, or would put enormous strain on the neck, were it not for a unique adaptation of the hare's skull. An animal's cranium, or skull, comprises several bone plates. The bones are loosely connected at birth. This allows for an easy birth, but the bones fuse firmly soon after birth. Hares, however, retain a soft connection at the back of the skull throughout their life. This connection cushions the back of the head when the animal bounds quickly.

Hares and rabbits are the only placental mammals to have an articulated (jointed) skull. Other mammals, including hares' extinct ancestors, have a completely fused cranium. Reptiles, amphibians, and some fish have a similarly articulated skull.

Hares have a flexible spine that enables a powerful kick from the long and sturdy hind leg bones. These are attached to a large pelvis. The pelvis restricts the leg movement mostly to back and forth, maximizing the power of the forward thrust. The back feet are also much longer than the front, with longer toe bones. Rabbits have shorter back legs than hares, although rabbits' back legs are still sizable.

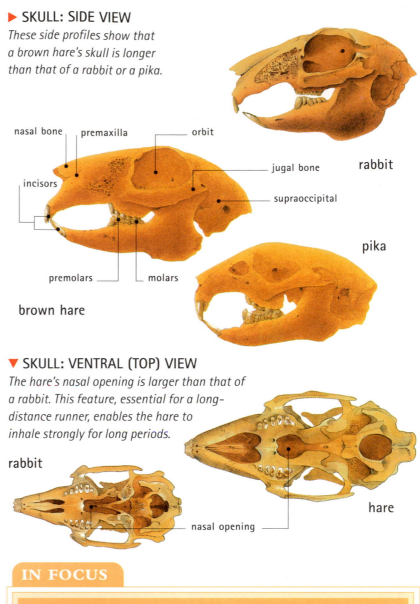

▶ SKULL: SIDE VIEW
These side profiles show that a brown hare's skull is longer than that of a rabbit or a pika.

nasal bone   premaxilla   orbit   rabbit

incisors   jugal bone

supraoccipital

premolars   molars   pika

brown hare

▼ SKULL: VENTRAL (TOP) VIEW
The hare's nasal opening is larger than that of a rabbit. This feature, essential for a long-distance runner, enables the hare to inhale strongly for long periods.

rabbit

nasal opening   hare

IN FOCUS

Postures

Hares and rabbits hold the carpal and metacarpal bones of their front feet in a vertical position at all times. They stand with their forelegs permanently on tiptoe, a posture called digitigrade. The digitigrade posture is ideal for fast running, and the back legs also hold a digitigrade posture when the hare is moving quickly; the tarsals and metatarsals are held upright. When a hare is bounding slowly, at rest, or standing upright on its back legs, its back feet rest flat on the ground. That is called a plantigrade posture.

hare's hind leg

The metatarsals are raised off the ground. This is called the digitigrade posture.

rabbit's hind leg

Muscular system

CONNECTIONS

COMPARE the massive biceps femoris and semitendinosus muscles in a hare's back legs with those of the back legs of a **CROCODILE** or a **GRIZZLY BEAR**.

The brown hare is one of the fastest animals on Earth. It can run in a straight line at 50 miles per hour (80 km/h), the same as the speed limit for cars in some states. Some reports suggest hares may be able to run even faster. The brown hare needs to be fast because it is a key prey species for many predators.

The brown hare's forelegs are not heavily muscled. They act as struts to support the weight of the animal, either at rest or when running. A running hare keeps its front legs straight, so there is no need for a bulky array of muscles. The forward thrust in a hare's bound comes almost entirely from the powerful back legs. The pelvis attaches to the

IN FOCUS

Chewing

Lagomorphs chew using their cheek muscles. The most enlarged are the pterygoids. The masseters are not large and the temporalis muscle is small. Together, these cheek muscles power the chewing action. They attach from the snout section of the skull to the lower mandible (jawbone) and power vertical and sideways movement during chewing.

▶ **SUPERFICIAL MUSCLES**
Brown hare
Although the basic muscle plan is similar to that of other mammals, the hare is characterized by very powerful muscles in its hind legs.

The biceps femoris is a large and powerful muscle that attaches the fibula bone to the pelvis. Along with other muscles, such as the semitendinosus, it powers the back legs.

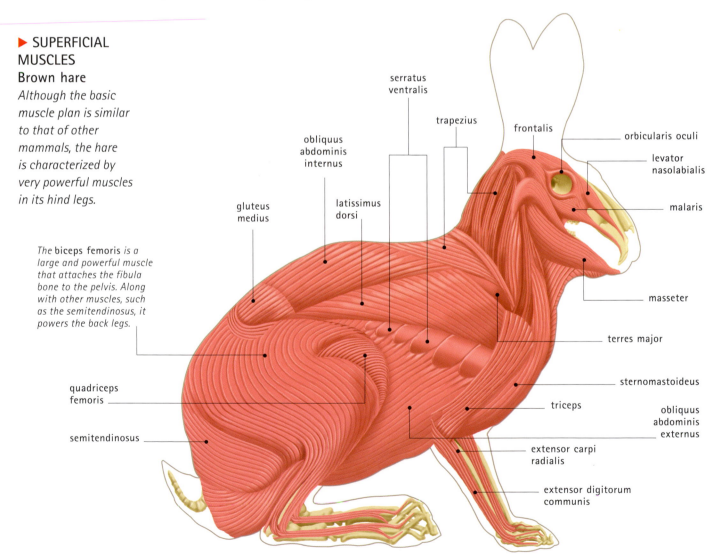

serratus ventralis
trapezius
frontalis
orbicularis oculi
levator nasolabialis
obliquus abdominis internus
malaris
latissimus dorsi
gluteus medius
masseter
terres major
sternomastoideus
triceps
obliquus abdominis externus
quadriceps femoris
extensor carpi radialis
semitendinosus
extensor digitorum communis

thigh by a number of large, strong muscles. These are the muscles that thrust a hare forward at great speed. The biceps femoris muscle attaches to the fibula bone, covering the semitendinosus, gluteus superficialis, and other large muscles that attach to the femur. The lower part of the rear leg is less heavily muscled. It provides leverage and balance rather than power.

Locomotion

Animals that run with a full bounding gait have large and heavily muscled back legs that provide power and forward thrust. Their shorter, less muscled forelegs land together and take the weight of the animal as it leans forward and draws its hind legs under the body for the next bound.

Hares have a distinctive lolloping gait. When walking slowly, a hare extends its long hind legs and leans forward onto its shorter forelegs one at a time, raising the back to bring the back legs forward together. The motion looks ungainly at slow speed but becomes more

▼ This speeding hare is driven forward by the powerful biceps femoris muscles of its hind legs.

fluid the faster the hare runs. Hares run using a gait called a half bound. A half bound also takes its thrust from the muscles of the back legs and balances on the front limbs.

Hares have pads on their feet, with rough hairs growing between them. These give grip, much like the tread on a shoe or tire. A firm grip keeps the hare from slipping. It also increases the hare's speed by allowing the muscles to transfer more power when pushing against the ground.

CLOSE-UP

The rough with the smooth

Mammals have three types of muscles: striated, smooth, and cardiac. Smooth muscle is named for its appearance; striated, or rough, muscle has a striped appearance. Organs such as the stomach are powered by smooth muscle. Smooth muscle is slow but efficient: 50 times slower but 300 times more energy efficient than striated muscle. Striated muscle, such as that in the muscles of a hare's legs, is quicker than smooth muscle but uses far more energy. Striated muscle moves the limbs. Tendons anchor striated muscles to the bones; smooth muscles have no tendons. The heart is powered by cardiac muscle.

Nervous system

CONNECTIONS

COMPARE the small brain of a hare, which weighs just 0.35 to 0.45 ounce (10 to 13 g), with the much larger brain of a *DOLPHIN*, an *ELEPHANT*, or a *GRAY WHALE*.

All mammals have a spinal cord at the center of their nervous system. Impulses pass through nerves in the spinal cord from the brain to the body. The delicate and irreparable nerves of the spinal cord are cushioned from physical harm by a liquid called cerebrospinal fluid. A layer of membranes encases the cerebrospinal fluid. The liquid-filled cushion containing the delicate nerves of the spinal cord fits inside the vertebrae, protected by the surrounding bone, muscles, and layers of fat. Nerves in the spinal cord carry messages along the back to different parts of the body. At each vertebra, paired nerves branch off the spinal column and spread around the body. One of the pair carries messages to the muscles; the other carries sensory signals back to the brain.

Good sense

Lagomorphs use a number of external senses during their daily life. A hare's vision extends almost 360 degrees around its head. Its eyes are large and gather a lot of light. The eye is more sensitive at night than a human eye. Many lagomorphs have nocturnal or crepuscular (dawn and dusk) habits. A large nerve called the optic nerve carries signals from the light-sensitive retina (inside the lining of the eye) to the optical cortex at the rear of the brain.

IN FOCUS

Lagomorph senses

Hares have sensitive eyesight, hearing, and sense of smell. They need to be alert to predators constantly, so they sniff the air and rotate their ears in vigilance. Hares also use odor to test their food. The olfactory, or smelling, cells lie in the snout. Hares have a touch-sensitive pad under a skin fold near the opening of their nostrils. Lagomorphs leave scent messages, which are secreted in liquid form from scent glands on their face and around the anus.

Hares and rabbits are constantly alert. They look, listen, and sniff for potential danger.

▶ **Brown hare**
This schematic diagram shows the main elements of the central nervous system (brain and spinal cord) and the peripheral nervous system. Not all nerves are depicted.

The **brain** *weighs 0.35 to 0.45 ounce (10–13 g). Nerves pass directly from the lower brain to the eyes, nose, face, and heart.*

cranial nerves

right cerebral hemisphere

cerebellum

spinal cord

lumbar plexus

mandibular nerve

median nerve

sciatic nerve

Circulatory and respiratory systems

Blood is pumped around the body by the heart. It is usual for smaller mammals to have a higher heart rate than larger ones. A hare's heart beats between 150 and 300 times a minute. The average is 200 to 240 times, or three or four times the human heart rate.

The mammalian heart has four chambers. Two chambers are called atria, and two are ventricles. The ventricles are larger. The heart sends blood first to the lungs to pick up oxygen and lose carbon dioxide. Cells use oxygen to liberate the energy needed to carry out their functions. The activity of the cells, in turn, allows the body to function. The oxygenated (oxygen-rich) blood returns to the larger left side of the heart and is then pumped to the body. A lagomorph's heart sits centrally in the chest cavity. In humans, it lies slightly to the left.

Arteries are elastic blood vessels that carry blood away from the heart and serve the major organs. Arteries expand and constrict and help the heart force the blood quickly to the major organs. Oxygen-depleted blood returns to the heart through less elastic vessels called veins. Large veins have valves that prevent the blood from flowing backward. As the hare moves, the muscles that surround its veins help squeeze the blood back toward the heart.

▼ **Brown hare**
The red vessels carry oxygen-rich blood from the lungs to other parts of the body; the blue vessels carry oxygen-depleted blood back to the lungs.

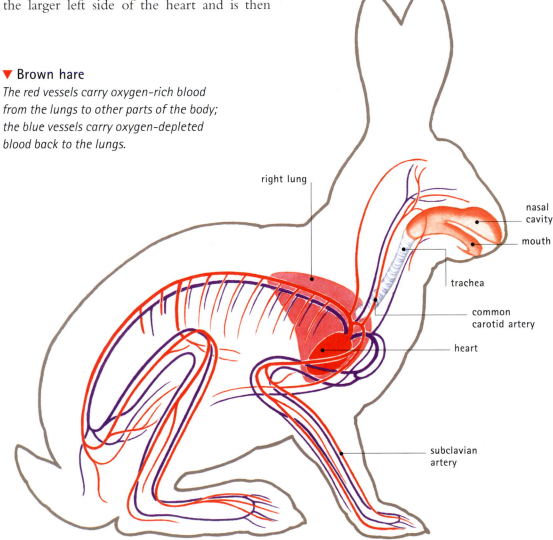

right lung

nasal cavity

mouth

trachea

common carotid artery

heart

subclavian artery

Blood disease

Viral hemorrhagic disease (VHD) is a serious condition in hares and rabbits that causes severe bleeding. VHD infection is caused by a virus. Rabbits with VHD suffer from uncontrollable bleeding in their gut. VHD has a short incubation period of 1 or 2 days. When a rabbit becomes infected with VHD it appears healthy for the incubation period. Ninety percent of rabbits that fall ill with VHD die within 5 days. Young rabbits recover more often than adults. The internal bleeding in a rabbit with VHD often shows as blood in the feces. By the time the victim dies, it often has a liver that is falling apart (friable) and a lower gut filled with foul-smelling liquid caused by internal bleeding and undigested food. The dead animal's stomach is often full of food that it could not digest because of the disease. Rabbits can catch VHD from other rabbits. They also fall prey to the condition during periods of stress or when they have a change in food supply.

The lymphatic system

Mammals have two circulation systems: the blood system and the lymphatic system. The former carries oxygen and carbon dioxide around the body in red blood cells. Blood also absorbs and carries the products of digestion from the gut to an animal's tissues and organs. As well as red blood cells, the blood also contains white blood cells. The white cells are called lymphocytes. They protect the animal by killing infectious organisms and dismantling dead tissues. Some lymphocytes move around and look similar to amoebas. Lymphocytes are present in the blood, but the blood system is not the body's main way of transporting lymphocytes. They also move through the body's second circulatory system, the lymphatic system. Lymph is a clear body fluid that circulates around the body through the lymphatic system, which works in parallel to the blood system. Lymph contains many lymphocytes. They move in and out of the lymphatic system and into the blood and tissues wherever they are needed.

Myxomatosis

Myxomatosis is a naturally occurring infectious viral disease of rabbits in South America. The disease affects hares far less frequently. Blood-feeding insects such as mosquitoes and fleas spread the virus. Infected rabbits develop swellings under the skin around the eyes and ears. The eyes weep and swell up with lymph and pus until the animal cannot see. The rabbit's lymph system becomes overloaded by myxomatosis, and the animal dies after around two weeks. People have used the disease as a way of controlling rabbit numbers in Australia and Europe. Populations that are newly exposed to myxomatosis are highly vulnerable, but over time they may become more able to resist the infection—and their numbers recover. Myxomatosis is a controversial form of biological control (using organisms to control other organisms) of rabbits because of the suffering it causes.

▲ *Its swollen eyes indicate that this rabbit has myxomatosis.*

Digestive and excretory systems

Hares and rabbits are almost exclusively vegetarian. Arctic hares sometimes eat carrion and often raid baited traps left for foxes. Otherwise, lagomorphs eat mostly grasses and herbs. They also nibble at buds, bark, and twigs, particularly at times when grasses are hard to find. Grasses contain sugars and proteins but can be digested only slowly. Lagomorphs have a digestive system that is able to process a large amount of food in a relatively small space. Hares eat a lot: two or three hares consume as much grass as one sheep despite being much smaller.

After it is swallowed, food travels into the stomach. The newly swallowed food mixes with the stomach contents and passes down the gut, or intestine. The intestine is screw-shaped, like a spiral stairway. The shape maximizes both the distance to the hindgut,

CLOSE-UP

B vitamins

Of all the vitamins essential to life, the B complex, and B_{12} in particular, is hardest for a plant-eating animal to include in its diet. A deficiency of B_{12} impairs an animal's ability to synthesize DNA and interferes with the production of red blood cells. Vitamin B_{12} is unlike some other vitamins, such as vitamin D, in that animals cannot produce their own B_{12}. It is available directly only from bacteria, fungi, and algae. Many animals obtain B_{12} from these foods and store it in the liver. Carnivores get their B_{12} from the flesh of animals that have obtained B_{12} from the primary sources. Humans also get B_{12} from other foods, such as eggs, dairy products, and yeast products. None of these sources of vitamin B_{12} is available to hares. The bacteria in the hare's cecum make B_{12}, but they do so slowly. By double-digesting their food, hares enable the bacteria in the cecum to make 0.1 milligram of B_{12} for every kilogram of food they eat. The bacteria also make vitamins B_1 and K.

▶ **Brown hare**

The digestive system is unusual, since food is digested twice. The products of the first stage of digestion (cecal pellets) are expelled and immediately swallowed. Double-digestion ensures that the hare derives maximum benefit from its food.

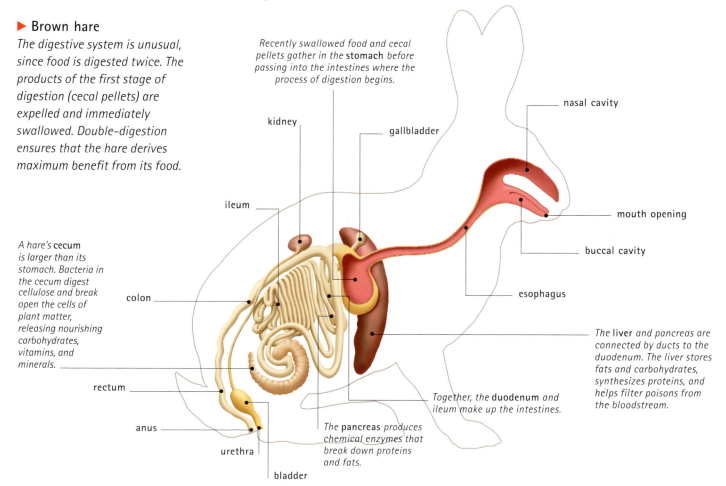

*Recently swallowed food and cecal pellets gather in the **stomach** before passing into the intestines where the process of digestion begins.*

kidney

nasal cavity

gallbladder

ileum

mouth opening

*A hare's **cecum** is larger than its stomach. Bacteria in the cecum digest cellulose and break open the cells of plant matter, releasing nourishing carbohydrates, vitamins, and minerals.*

buccal cavity

colon

esophagus

rectum

*The **liver** and pancreas are connected by ducts to the duodenum. The liver stores fats and carbohydrates, synthesizes proteins, and helps filter poisons from the bloodstream.*

anus

*Together, the **duodenum** and ileum make up the intestines.*

urethra

*The **pancreas** produces chemical enzymes that break down proteins and fats.*

bladder

▲ *Sometimes, a brown hare will stand on its hind legs to reach food, as this individual is doing to eat a corncob. Usually, however, brown hares eat grass and other plant matter from the ground.*

allowing time for lots of digestion, and contact with the surface area for absorbing the nutrients into the blood. In the hindgut the food enters an organ called the cecum, which is an offshoot of the gut. Hares and other lagomorphs have a very large cecum, where most digestion takes place. A hare's cecum is 10 times the size of its stomach. The cecum is a blind alley in the hindgut where food gathers.

Bacteria

An array of beneficial bacteria in the cecum digests cellulose and breaks open the plant cells, releasing the carbohydrates, vitamins, and minerals that nourish the hare. Lagomorphs cannot digest food without the bacteria in their cecum. Antibiotics that cure human diseases can harm a lagomorph because they may kill the bacteria and prevent the animal from digesting its food.

The vital ingredients in the food pass through the cecum wall and directly into the hare's blood system. After digesting for a time in the cecum, the food forms into a type of feces called cecal pellets. The cecal pellets pass out of the anus, like typical feces, but they are not waste. The hare has not yet finished its food—it eats the soft, green cecal pellets, swallowing them after they emerge from the anus.

Initially, the cecal pellets gather in the stomach. Then, partly digested food passes back up from the gut into the stomach and mixes with the pellets. The mixture passes back into the gut for another round of digestion. Lagomorphs keep some food from

CLOSE-UP

Food requirements

European rabbits eat the equivalent of 5 percent of their body weight in food (dry weight) each day. They drink between 8 and 10 percent of their body weight in water every day. A growing or pregnant rabbit requires 15 to 20 percent protein in its food; at other times of life 12 to 15 percent protein will do. It takes 4 to 5 hours for food to pass through a rabbit but a laboratory rabbit that has been denied food will not empty its stomach completely for 9 days.

their last meal in their stomach and mix it with the next. Mixing accelerates digestion of the new meal because it exposes food to cellulose-digesting bacteria from the cecum. It takes only three to four hours for food to pass through a hare, but because of mixing it is more than a week before every last bit passes through.

Young hares and rabbits do not need bacteria to digest milk. However, they need to build up a population of bacteria within their cecum for the time when they eat solid food. Young lagomorphs begin to introduce some solids into their diet while they are still drinking milk. This gradual change in diet enables the necessary bacteria to multiply in the lagomorphs' digestive system. The move to solid foods is called weaning. Weaning lagomorphs probably get their bacteria from eating their mother's cecal pellets.

▲ The grass that this hare is eating will pass through its intestines twice. This process ensures that the animal extracts the maximum benefit from its food.

IN FOCUS

Two guts for the price of one: Coprophagy

The eating of cecal pellets, or coprophagy, is useful to a hares and rabbits in many ways. By digesting plants twice, a process called refection, lagomorphs can take in far more nutrients from their food than they could from a single pass. Bacteria living in the cecum digest the hare's food, but the cecum is far back in the gut and the bacteria have a time limit for their service. By eating the food a second time, the hare gives the bacteria more time to break down the food. Most important of all, double-digestion is the only way a hare can obtain its full quota of vital vitamins.

Hares produce cecal pellets when they are resting and true feces during the active part of the day. Hares that feed by day produce cecal pellets at night and feces in the day; nocturnal (night-active) lagomorphs do the reverse. Feces are waste and need to be disposed of away from the animal's sleeping area for the sake of hygiene. Dropping feces while foraging scatters them at a hygienic distance. Communal rabbits use latrines away from their burrows. Producing feces at different times from cecal pellets also has hygiene benefits. It is convenient for a hare to produce and eat cecal pellets while resting.

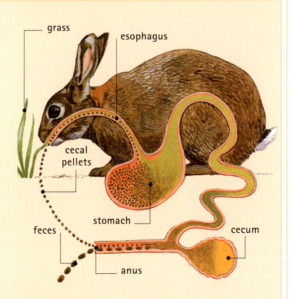

CECAL PELLETS AND FECES
Cecal pellets, or cecotrophs, are soft, green, and covered in mucus. True feces, however, are hard, black, and roughly spherical.

Cecal pellets are a guaranteed food source at a time when the animal would be unable to eat otherwise. Coprophagy allows hares to digest their food more evenly through the day.

Reproductive system

CONNECTIONS

COMPARE the uterus of a hare with that of a *CHIMPANZEE*. That of the hare has two horns and two cervices, while the chimpanzee's has only one of each.

COMPARE the gestation period of a hare with that of an *ELEPHANT*.

Studies of hare behavior in Scotland have revealed that hares breed throughout most of the year, with a peak in pregnancies occurring in April and May. The length of daylight is an important trigger for breeding activity. Each year, a female has an average of three litters of (usually) four young; the young are called leverets. Early and late litters contain fewer, and smaller, leverets than spring litters. For fertilization to occur, a female mammal needs to release an egg from one of her ovaries soon before or after a mating. Many animals ovulate at regular intervals. Female hares maximize their chances of a successful mating by ovulating in response to mating.

Rapid growth

The length of pregnancy, or gestation period, of a brown hare is 41 or 42 days. At birth, leverets weigh 3.8 to 5.3 ounces (110–150 g). Suckled on rich, fatty milk from their mother's six mammary glands, they grow quickly. They may double their body weight within three or four weeks, and after about five weeks they reach their adult weight of 5.5 to 15 pounds (2.5–7 kg). Male brown hares are 5 percent heavier than the females. Males mature at around six months and females at around seven or eight months.

Lagomorphs have a quick turnover of young, enabled by a double, or duplex, uterus. The duplex uterus has two horns and two cervices. The embryos develop on one side at a time. Up to a week before a female rabbit gives birth to a litter of kittens, or baby rabbits, she can mate and implant the next set of eggs on the second side of her uterus.

If feeding conditions are good she may be pregnant continuously while still suckling kits. If feeding conditions are poor, or if the female is under physical stress, she reabsorbs any embryos she is carrying. The rabbit is ready to mate again immediately. If conditions improve during the next pregnancy, she may nourish the embryos to full term. Even in laboratory conditions, around 20 percent of young rabbits die by the time they are weaned.

Male hares and rabbits have a unique arrangement of sexual organs. Lagomorphs are the only placental mammals with a scrotum that lies in front of the penis, toward the belly; in all others it hangs behind, toward the anus.

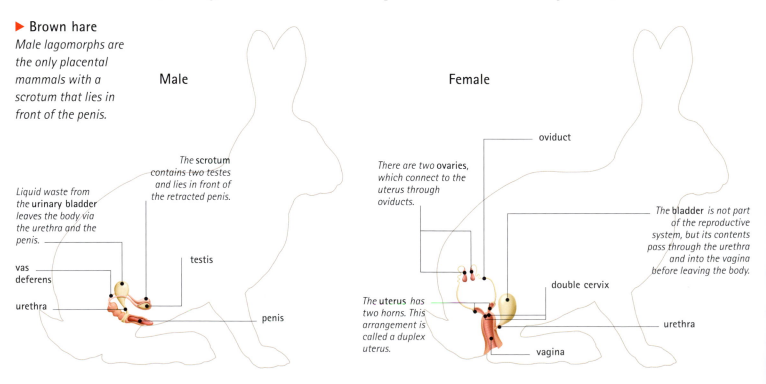

▶ **Brown hare**
Male lagomorphs are the only placental mammals with a scrotum that lies in front of the penis.

Male

Female

Liquid waste from the **urinary bladder** *leaves the body via the urethra and the penis.*

The **scrotum** *contains two testes and lies in front of the retracted penis.*

vas deferens

urethra

testis

penis

There are two **ovaries***, which connect to the uterus through oviducts.*

oviduct

The **bladder** *is not part of the reproductive system, but its contents pass through the urethra and into the vagina before leaving the body.*

double cervix

urethra

The **uterus** *has two horns. This arrangement is called a duplex uterus.*

vagina

This arrangement is probably an adaptation that protects the testicles from damage during half-bounding locomotion. After the breeding season, male hares' reproductive organs shrink. This temporary change keeps the sexual organs from getting in the way and protects them from harm at the times of year when they are not needed.

Altricial or precocious?

Different types of animals are born at different stages of development. A human baby is helpless at birth and relies on its parents for everything—food, warmth, protection, and grooming. It takes many years for humans to become independent. Humans are born altricial (dependent). Other animals never even meet their parents and are independent right from the start. These are precocious (independent) animals.

Baby rabbits are altricial at birth. They are hairless, blind, and unable to move around much because their muscles are not yet developed fully. Hares have precocious young that are fully furred and ready to run and hide, but even these rely on mother's milk for a time. Like all mammals, baby lagomorphs depend on their mother for milk to nourish them through their early growth and development. Mammals cannot be totally independent from birth.

GENETICS

Chromosomes

Chromosomes are structures in animal cells that are made of deoxyribonucleic acid, or DNA. The chromosomes are in the cell nucleus (control center). Before a cell divides, DNA winds up into denser chromosomes that stain and are easy to see with a microscope. Chromosomes come in pairs. Different animals have different numbers of chromosomes: rabbits have 44 pairs; hares have 36 to 46 pairs, depending on the species. Having different numbers of chromosomes makes breeding between hares and rabbits impossible: neither their chromosomes nor the genetic materials inside match.

IN FOCUS

Boxing hares

In late winter and early spring, adult male brown hares compete with each other to mate with sexually active females. Males that have already paired with females may chase away rival males. A female that is not ready to breed may also need to discourage the attentions of males. The rivals rear up on their hind legs and hit out at each other with their forepaws. A struggle akin to a boxing match follows. The expression "mad March hares" derives from this behavior, since the squabbles are typical of the early part of the breeding season, especially in March.

An animal's lifestyle reflects whether or not it has precocious young. Rabbits dig a burrow for their young. Inside it is safe, warm, and dry. The mother can make a nest from dry grasses and her own soft underfur in which she can brood her litter. Pregnancy is a risky time for any animal. A heavily pregnant rabbit suffers from reduced mobility and might easily fall prey to a fox.

Brown hares live above ground, and the leverets are born in a small nest that cannot hide and protect a naked, blind, helpless baby. Leverets are born fully furred and able to run. Newborn leverets scatter and hide among plants, coming out of hiding only once a day to suckle from the female. Suckling is risky for the whole family, and there is no time to waste; a mother brown hare can produce enough milk for the whole litter in just five minutes. It is their only feed of the day, and their only contact with either parent. During the peak period of the breeding season, when the best food is available to the female, leverets move on to solid food after around 23 days. Late in the season, when the supply of food available to the mother is not so good, they may suckle for three months.

JOHN JACKSON

FURTHER READING AND RESEARCH
Kardong, Kenneth V. 1995. *Vertebrates.* William C. Brown Publishers: Dubuque, IA.
Macdonald, David. 2006. *The Encyclopedia of Mammals.* Facts On File: New York.
Walker's Mammals:
  http://www.press.jhu.edu/books/walkers_mammals_of_the_world/lagomorpha/html

Hawkmoth

PHYLUM: Arthropoda CLASS: Insecta FAMILY: Sphingidae

Hawkmoths are a family of around 1,000 species of large, fast-flying moths. Hawkmoths occur worldwide except in the polar regions; the greatest number of species live in tropical regions. A few species are pests of crops.

Anatomy and taxonomy
Scientists seek to classify all organisms into related groups, based largely on anatomical features. Hawkmoths are members of a large group, or order, of insects called Lepidoptera, which includes all butterflies and moths.

● **Animals** Animals are multicellular and rely on other organisms for food. They differ from other multicellular life-forms in their ability to move around (generally using muscles) and respond rapidly to stimuli.

● **Arthropods** In terms of number of species, arthropods are by far the biggest group of animals on Earth. They include crustaceans (crabs, shrimp, and relatives), arachnids such as spiders and scorpions, myriapods such as millipedes and centipedes, and insects. Arthropods are invertebrates (animals without backbones) that have jointed limbs. Their body has a hard outer layer, which protects them from damage and acts as an anchorage for the muscles. Arthropod blood, or hemolymph, is not channeled in vessels. Instead, it fills much of their internal space, bathing the internal organs and providing them with nutrients. Arthropod limbs may act as legs, as gills, or as swimming structures, depending on the species. Some arthropods have evolved to lose their limbs entirely.

● **Uniramians** The word *uniramian* means "one branch," and refers to the one-branched limbs of these arthropods. This contrasts with crabs and their relatives; these "biramous" arthropods have two-branched limbs. Uniramians are mainly land-living arthropods; they include myriapods as well as insects. They typically have a pair of mandibles although these are absent in some species. Uniramians breathe using air tubes called tracheae, which lead into their body from small openings in their sides.

● **Hexapods** Hexapods are six-legged arthropods. The vast majority of hexapods are insects, but there are three small noninsect hexapod groups. Noninsect hexapods, such as springtails, do not have wings or antennae; they are further

▶ *This family tree shows the butterflies, skippers, and one of the 20 moth superfamilies. Approximately 150,000 species of lepidopterans have been described, of which more than 130,000 are moths.*

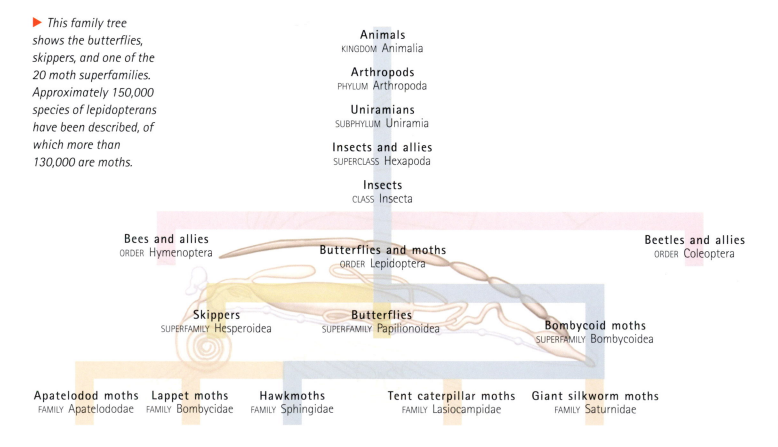

Animals
KINGDOM Animalia

Arthropods
PHYLUM Arthropoda

Uniramians
SUBPHYLUM Uniramia

Insects and allies
SUPERCLASS Hexapoda

Insects
CLASS Insecta

Bees and allies
ORDER Hymenoptera

Butterflies and moths
ORDER Lepidoptera

Beetles and allies
ORDER Coleoptera

Skippers
SUPERFAMILY Hesperoidea

Butterflies
SUPERFAMILY Papilionoidea

Bombycoid moths
SUPERFAMILY Bombycoidea

Apatelodod moths
FAMILY Apatelododae

Lappet moths
FAMILY Bombycidae

Hawkmoths
FAMILY Sphingidae

Tent caterpillar moths
FAMILY Lasiocampidae

Giant silkworm moths
FAMILY Saturnidae

separated from the insects by the structure of their mouthparts, which are kept in a pouch on the underside of the head and are popped out for feeding.

● **Insects** These animals are by far the largest group of arthropods. Adult insects have a body divided into a head, thorax, and abdomen, three pairs of legs, and compound eyes composed of many separate facets. Most adult insects can fly, a major factor contributing to their success.

● **Endopterygotes** The most successful insect groups are the beetles, butterflies and moths, flies, and hymenopterans (wasps, bees, and ants). Along with a few smaller groups, these insects are referred to as endopterygotes. Adult and larval (young) endopterygotes are very different structurally. There is an inactive stage called the pupa, during which most of the larva's tissues are broken down and the adult structures are built up.

● **Butterflies and moths** Butterflies and moths form the insect order Lepidoptera, or "scale wings"; lepidopteran wings are typically covered by tiny flat scales. These scales are often brightly colored. Young lepidopterans, or caterpillars, feed with biting mouthparts, but nearly all adults are without such mouthparts. Instead, they sip sweet liquids like nectar through a long tube called a proboscis. Lepidopterans are divided into a number of groups called superfamilies. For example, butterflies belong to one or the other of two closely related superfamilies. Members of all the other superfamilies are called moths.

● **Butterflies** The superfamily Papilionoidea contains the 20,000 or so species of "true" butterflies. In contrast with most moths, butterflies are day-flying and usually fold their wings vertically behind their back at rest. Their pupae, called chrysalises, usually lack a protective cocoon and are formed above ground. The small skipper butterflies are classified in a separate superfamily, Hesperoidea.

▲ At rest, the poplar hawkmoth adopts an unusual position: it holds its hind wings in front of the forewings.

● **Bombycoid moths** This group includes several families of large-bodied moths. As well as hawkmoths, it includes the silkworm moths—among them the giant atlas moths of Asia, the largest moths in the world.

● **Hawkmoths** The hawkmoths form the family Sphingidae. It is named for the fact that some caterpillars, when disturbed, raise up their front end so that their shape is a little like the stone-carved Sphinx of ancient Egypt. Adult hawkmoths have long, narrow forewings, and a bulky but streamlined body. The caterpillars usually have a hornlike structure growing near their back end and are called hornworms. Some biologists place hawkmoths in a superfamily of their own, Sphingoidea.

FEATURED SYSTEMS

EXTERNAL ANATOMY Adults have wings and parts of the body covered with tiny, often brightly colored scales. Adults suck liquid food through a coilable tube called the proboscis. Caterpillars are soft-bodied and have extra non-jointed legs to grip onto surfaces. *See pages 496–499.*

INTERNAL ANATOMY Most internal organs are suspended in an open, hemolymph-filled cavity called the hemocoel. An adult's digestive system is quite different from a caterpillar's. *See pages 500–501.*

NERVOUS SYSTEM This consists of a brain and separate nerve centers called ganglia. They are connected by a nerve cord, with individual nerves arising from the ganglia. The nervous system is added to, rather than re-created, during metamorphosis. *See page 502.*

CIRCULATORY AND RESPIRATORY SYSTEMS Hemolymph is pumped between the thorax and abdomen with the aid of a dorsal vessel, or heart. Gas exchange takes place through a separate system of air tubes. *See page 503.*

REPRODUCTIVE SYSTEM As in most animals, males produce sperm, and females produce egg cells. There is an active caterpillar stage before an inactive pupal stage, during which adult features develop. *See pages 504–505.*

External anatomy

CONNECTIONS

COMPARE a hawkmoth's compound eyes with the eyes of an *EAGLE*. The hawkmoth has compound eyes made up of many separate units; the eagle's eye uses a single lens to focus an image onto light-sensitive cells.

COMPARE the antennae of a hawkmoth with those of a *LOBSTER*. Both are used for detecting odors.

The body of an adult moth, like that of any other insect, is broadly divided into three sections: head, thorax, and abdomen. The wings and legs arise from the thorax. All three sections are surrounded by a tough outer covering, the exoskeleton, made mainly of a substance called chitin. The exoskeleton consists of separate hard plates called sclerites, connected by a thinner cuticle that provides the insect with flexibility. In lepidopterans, the cuticle is often hidden by the covering of scales and hairs. The exoskeleton protects the internal organs, helps resist water loss, and serves as an attachment for many of the hawkmoth's muscles.

Hawkmoth heads

A hawkmoth's head contains the main organs and structures for sensing its environment. Prominent are two large eyes, called compound eyes, which have a hard outer covering like the rest of the body. Compound eyes are a feature of most adult insects and are made up of many tiny separate eyes fitted together into a mosaic pattern. Hawkmoth eyes are different structurally from those of most other insects, since they possess various adaptations for vision in low-light conditions.

Insect antennae are sometimes called "feelers," but moths do not generally use them much for touch. Moths' feathery antennae are vital for detecting odors in the air, and moths also use their antennae to measure their airspeed in flight, by detecting how much they are bending as the air flows past. Butterfly antennae are different from those of moths; they are long and slender, with a clublike swelling at the tip.

Mouthparts

The mouthparts of adult lepidopterans are specialized for sucking up liquids. Except in a few species of unusual moths, these insects do

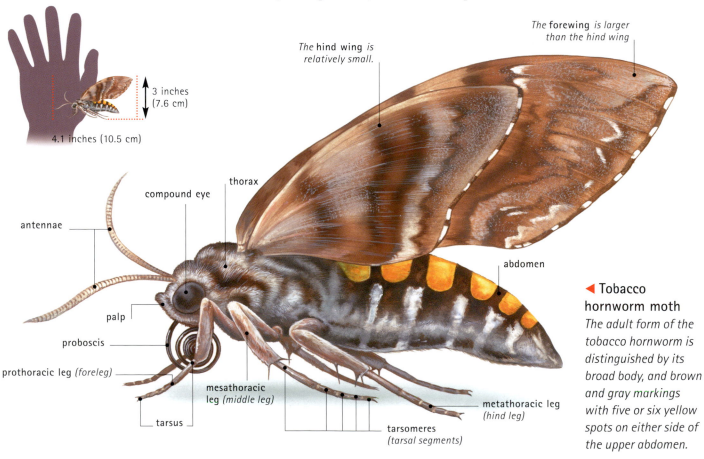

3 inches (7.6 cm)

4.1 inches (10.5 cm)

The **hind wing** *is relatively small.*

The **forewing** *is larger than the hind wing*

antennae

compound eye

thorax

palp

proboscis

prothoracic leg *(foreleg)*

tarsus

mesathoracic leg *(middle leg)*

tarsomeres *(tarsal segments)*

metathoracic leg *(hind leg)*

abdomen

◄ **Tobacco hornworm moth**
The adult form of the tobacco hornworm is distinguished by its broad body, and brown and gray markings with five or six yellow spots on either side of the upper abdomen.

IN FOCUS

Sensitive sensilla

Insects have microscopic structures called sensilla all over their bodies. Sensilla are used for detecting chemicals, and some are touch-sensitive. Antennae are the main organs of smell in moths and butterflies. A male tobacco hornworm moth has around 400,000 odor-detecting sensilla on each antenna. About three-quarters of these are targeted to detect specific sex-attractant chemicals, or pheromones, which are released by the female moth. The extraordinary sensitivity of the antennal sensilla allows male moths to find mates from a very great distance. By contrast, sensilla on a female's antennae are used mainly to detect plant odors. This ability helps her find the right plant to lay her eggs on. Other sensilla are used for tasting chemicals. For example, some butterflies taste food with sensilla on their feet.

The compound eye is made of thousands of individual cells called ommatidia.

labial palps

A hawkmoth's antenna is covered with microscopic sensilla that are used for detecting odors.

The proboscis is in a coiled position.

◄ HEAD
The hawkmoth's long proboscis is not always apparent, since it is kept coiled up when not in use.

not have biting mouthparts. Instead, each of the two maxillae forms one half of a long, flexible tube called the proboscis. The halves are bound together by a zipperlike mechanism. The moth uses its proboscis like a drinking straw, usually to sip nectar from flowers. Some moths also use the proboscis to suck up liquid in damp sand or mud. This behavior, called puddling, allows the moth access to minerals and salt ions not present in nectar. Other liquids can also be drunk, such as fermenting fruit pulp; vampire moths use their proboscis to suck blood like a mosquito. The proboscis is straightened by increasing the pressure of the hemolymph. It is coiled up under the head when not in use.

CLOSE-UP

Moth proboscises

Hawkmoths have some of the longest proboscises of all. They use them for feeding from deep, narrow flowers. The record for the longest proboscis is held by a hawkmoth from Madagascar; it has an 11-inch (28-cm) proboscis that it uses to drink nectar from star orchids. Some species of moths have nonfunctional mouthparts. These species survive on food reserves they stored up as caterpillars.

A colored scanning electron micrograph showing the tip of a hummingbird hawkmoth's proboscis (magnified 500 times).

▼ **FEEDING**
Hummingbird hawkmoth
A hummingbird hawkmoth hovers in front of a flower and extends its very long proboscis deep into the flower's nectary.

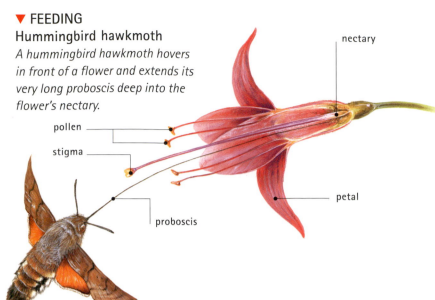

nectary

pollen

stigma

petal

proboscis

Body and legs
Behind the head, the body is divided externally into a number of segments. The thorax is formed by three segments, and 10 more form the abdomen. The thorax contains the powerful wing muscles. The legs are joint–ed and end in claws that are important for clinging to surfaces. In many butterflies, the first pair of legs is not used for walking; these legs are instead important for tasting and are held against the body when not in use.

Many moths have drumlike hearing organs on their thorax or abdomen. They allow the moths to hear ultrasonic squeaks produced by echolocating bats. This alerts the moths to take urgent evasive action. Hawkmoths lack such organs, but some groups can hear ultrasound through unique hearing organs. Two different hawkmoth groups have ears that are actually highly modified mouthparts, formed from the labial palps and labral pilifers. Interestingly, these two types of ears appear to have evolved independently. They represent a remarkable example of the way that animals with different evolutionary histories can evolve structures that have similar functions. This phenomenon is called convergent evolution.

Eight pairs of breathing holes, or spiracles, open on the sides of the abdomen. Otherwise the abdomen has few external features, apart from the sex organs. A row of long scales at the back edge of each segment protects the softer cuticle underneath.

The wings
Insect wings are flat outgrowths of the body wall. Running across them are thick tubes called veins, which are hollow and enable

CLOSE-UP

Useful scales

Moth and butterfly wings are covered by thousands of tiny flat scales, arranged in rows like roof tiles. Scales are actually flat, hollow hairs. Often they are filled with natural chemical pigments, producing a range of colors. Many scales also have ridged surfaces that create metallic, iridescent colors when light falls on them. Scales easily come away from the moth's body. If a moth flies into a spiderweb, the scales prevent the web from contacting the body surface; by rolling downward the moth can escape from the web, shedding scales as it goes.

▶ *An individual moth scale is full of holes and is therefore very light. This construction reduces the weight of the scales and thus reduces the amount of energy a moth needs for flight.*

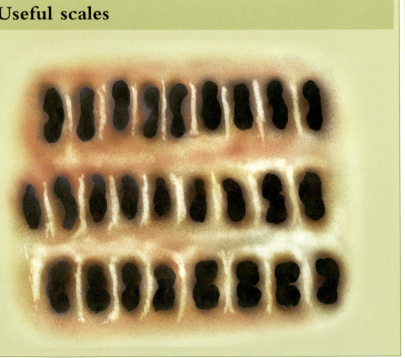

COMPARATIVE ANATOMY

Hairy or hairless

Hawkmoth caterpillars are almost hairless, but many other caterpillars are covered with long hairs. These are often irritating or contain poisonous chemicals, so they help defend against enemies. Like adult moths, caterpillars use color and pattern for a range of different purposes. Mostly they are used for camouflage or mimicry—there are caterpillars that mimic twigs and stems, bird droppings, and even small snakes; aston- ishingly, some caterpillars can mimic the smell of young ants, so worker ants take them back to their nests and care for them. Most hawkmoth caterpillars are greenish and camouflaged, but a few are brightly colored, warning potential predators how unpleasant they are to eat. Death's-head hawkmoth caterpillars can use their mandibles to make a clicking sound that deters predators.

mandibles · spiracles · horn · ocelli (simple eyes) · thoracic legs · abdominal prolegs · anal proleg

hemolymph, nerves, and air tubes to enter the wings. Veins also strengthen the wings.

In lepidopterans, the forewings and hind wings are coupled by hooklike mechanisms so they beat at the same time. Hawkmoths have long, narrow forewings and beat their wings faster than other large moths. This makes them powerful fliers. Some hawkmoths are also expert at hovering. Wing scales have allowed moths to develop colored wing patterns, which have various uses. Often they act as camouflage when the moth is resting. Bright patterns may indicate that a moth is poisonous; "eyespot" patterns may startle predators. Dark colors help absorb heat from the sun. Butterflies often identify potential mates by their color patterns.

The caterpillar

At first glance, a moth or butterfly caterpillar looks completely different from an adult, although on closer inspection they share many features. Hawkmoth caterpillars are large, impressive creatures, often several inches long when fully grown. They usually have a distinctive hornlike structure growing near their back end, giving them the alternative name hornworms. In newly hatched cater- pillars the horn can be almost as long as the rest of the body. However, the horn is not a stinger; its function remains a mystery.

A caterpillar's exoskeleton is much thinner and softer than that of an adult. Caterpillars also have only small, simple eyes (or ocelli), and their antennae are small, too. Unlike adults, they have powerful biting mouthparts called mandibles, which are absent in the adults. The mandibles are used to chew their plant food. Caterpillars also have a structure called a spinneret, through which they produce fine silken threads.

The division between thorax and abdomen is less obvious in a caterpillar than in an adult, although close study reveals that the legs on the thorax are jointed. The abdomen bears around five pairs of fleshy prolegs (false legs). These end in rows or circles of tiny hooks to help them grip plant surfaces.

Internal anatomy

CONNECTIONS

COMPARE the flight muscles of a hawkmoth with those of a **DRAGONFLY**, a **FRUIT BAT**, and a **HUMMINGBIRD**.

COMPARE the crop of a hawkmoth, which branches from the side of the gut, with the crop of a **HONEYBEE**, which is continuous with the gut.

Like other insects, the hawkmoth has an inside consisting of a fluid-filled space called a hemocoel in which its various internal organs lie. The fluid is the insect equivalent of blood; it is called hemolymph. The hemolymph provides food to the organs and removes waste products. It also contains cells that fight infection and act as a store of fluid. In caterpillars, up to 40 percent of the body weight can be hemolymph.

Food processors

The lepidopteran digestive system, or gut, runs from the mouth to the anus. The gut has three main sections: the foregut, midgut, and hindgut. At the front, muscular pumps in the mouth cavity and the front part of the foregut, the pharynx, allow the moth to suck up liquid foods such as nectar. The moth can store food in its crop, which—unlike that of most other insects—forms a side branch of the gut.

Food is absorbed in the midgut. The hindgut has various roles including preventing water loss and adjusting the levels of dissolved salts in the moth's body. The hindgut also helps process waste-laden fluids; they are emptied into the hindgut from excretory organs called Malpighian tubules.

Once absorbed through the walls of the midgut, food is stored and processed in tissues that are collectively called the fat body. In some hawkmoths the adults do not feed, and so their

▼ *The internal anatomy of a hawkmoth is similar to that of other insects, with the nerve cord running ventrally (close to the lower surface). This contrasts with vertebrates, in which the nerve cord runs dorsally (at the back or upper surface).*

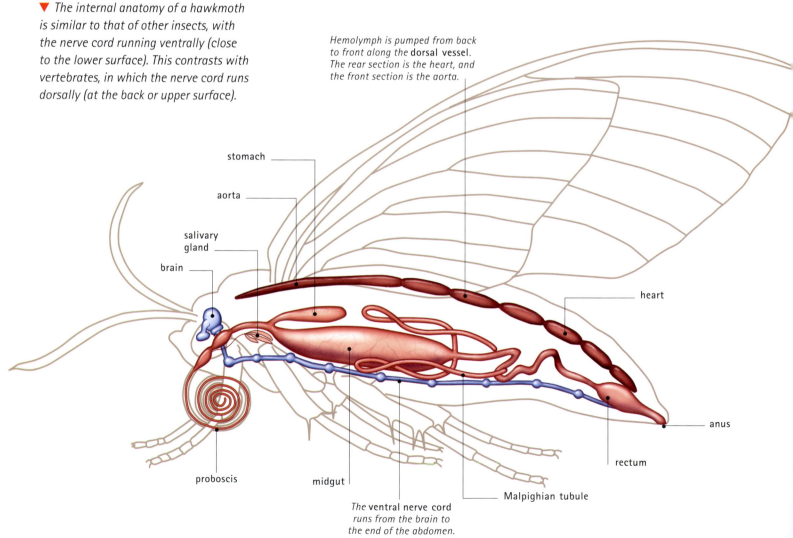

*Hemolymph is pumped from back to front along the **dorsal** vessel. The rear section is the heart, and the front section is the aorta.*

stomach

aorta

salivary gland

brain

heart

proboscis

midgut

anus

rectum

Malpighian tubule

*The **ventral nerve cord** runs from the brain to the end of the abdomen.*

CLOSE-UP

Powerful flight muscles

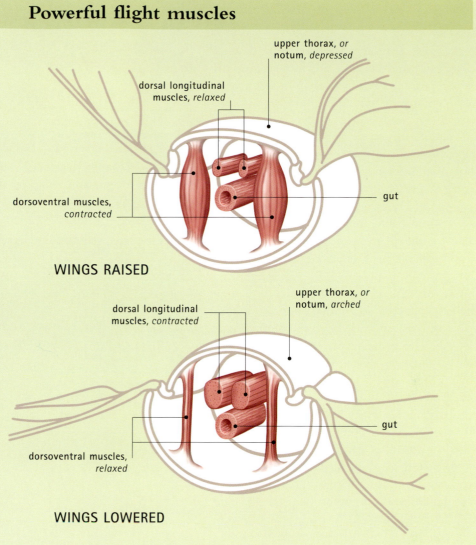

upper thorax, *or* notum, *depressed*

dorsal longitudinal muscles, *relaxed*

dorsoventral muscles, *contracted*

gut

WINGS RAISED

dorsal longitudinal muscles, *contracted*

upper thorax, *or* notum, *arched*

dorsoventral muscles, *relaxed*

gut

WINGS LOWERED

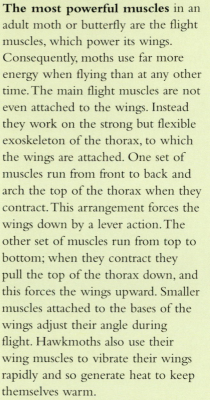

The most powerful muscles in an adult moth or butterfly are the flight muscles, which power its wings. Consequently, moths use far more energy when flying than at any other time. The main flight muscles are not even attached to the wings. Instead they work on the strong but flexible exoskeleton of the thorax, to which the wings are attached. One set of muscles run from front to back and arch the top of the thorax when they contract. This arrangement forces the wings down by a lever action. The other set of muscles run from top to bottom; when they contract they pull the top of the thorax down, and this forces the wings upward. Smaller muscles attached to the bases of the wings adjust their angle during flight. Hawkmoths also use their wing muscles to vibrate their wings rapidly and so generate heat to keep themselves warm.

▶ FLIGHT MUSCLES
In contrast to dragonfly wing muscles, none of a hawkmoth's main flight muscles are directly attached to the wings.

gut is nonfunctional. These moths rely on food supplies they built up when they were caterpillars and stored in their fat body.

All change

A caterpillar has a diet different from that of an adult moth and therefore has a different kind of digestive system. Nectar is almost entirely composed of sugar and water and needs little or no digestion before it can be absorbed. However, caterpillars feed on tough plant material that must be broken down and digested before it can be absorbed. The caterpillar digestive system has a large midgut to process food. It also produces a range of digestive enzymes to break down the food.

The muscles of caterpillars are also arranged in a pattern different from those of the adult. A caterpillar relies for movement on the pressure of the fluid inside its body, together with a complex arrangement of muscles in its body walls. By contracting these muscles it can reach out and change shape—a bit like squeezing a balloon. This type of structure is called a hydrostatic skeleton.

The caterpillar's internal organs are extensively reconstructed during the process of metamorphosis (the change from caterpillar to adult moth). Cells from the hemolymph help break down unwanted tissues. Nearly all of the muscles are digested before a new set of adult muscles begins to develop.

Nervous system

Like other insects, hawkmoths have a sophisticated nervous system that controls internal processes and helps them interact with their environment. Any nervous system consists of individual nerve cells called neurons. Neurons are specialized cells with long, thin, cablelike extensions called axons. They pass information to each other using electrical signals. Some neurons trigger muscle movements. Other neurons transmit incoming information from the eyes or other sense organs, while many more process the information that has been obtained.

Cords and ganglia

The basic structure of an insect nervous system is a nerve cord running down the underside of the body, with ganglia (concentrations of nerve cells) at intervals. Each ganglion is a semi-independent control center, although overall control rests with the insect's brain. Much of a moth's brain is devoted to processing visual information. Moth compound eyes are different from those of day-flying insects. Light is not restricted to a single unit but can move throughout the whole eye. This adaptation helps maximize the amount of light that can be detected, an essential adaptation for life in low-light conditions. Such eyes are called superposition eyes; a moth sees the world as a single image, as humans and other vertebrates do, albeit of low resolution.

► Along the length of the nerve cord are nodes of nerve cells called ganglia. Nerves branch from the ganglia to other parts of the hawkmoth's body. The nerve cord loops around the esophagus to reach the brain.

IN FOCUS

Rewiring

A caterpillar's nervous system is greatly reorganized when the animal begins to transform into an adult. Unlike the muscle system, however, it is not actually dissolved and broken down, but some regions such as the visual-processing part of the brain are expanded greatly. Several of the caterpillar's ganglia are drawn together and concentrated into a single structure in the adult.

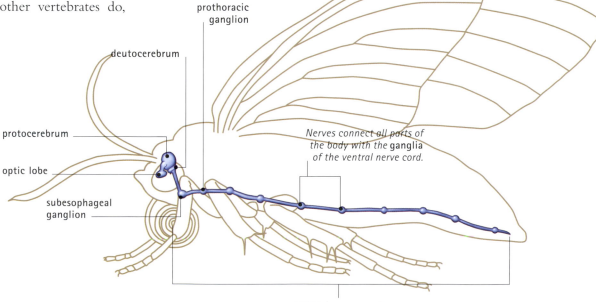

prothoracic ganglion

deutocerebrum

protocerebrum

optic lobe

subesophageal ganglion

*Nerves connect all parts of the body with the **ganglia** of the ventral nerve cord.*

ventral nerve cord

Circulatory and respiratory systems

When a human breathes, blood takes some of the oxygen from the air filling the lungs and transports it around the body. In insects, the relationship between the respiratory and circulatory systems is different. Hemolymph does not carry oxygen; instead, a separate system of air tubes delivers oxygen directly to the tissues.

Circulation

Insect hemolymph fills the main body cavity, or hemocoel, and has many functions. Almost the only enclosed vessel is the long, slender dorsal vessel. This muscular tube is able to pump the fluid inside it forward or backward. The wider back section, the heart, contains small valve-like openings, or ostia, connecting to the main body cavity. The narrower front section is the aorta.

In most insects, hemolymph is always pumped forward, toward the head. In lepidopterans, though (and also in beetles and flies), the direction of pumping is regularly reversed. This has the effect of causing hemolymph to ebb and flow between the thorax and abdomen, whose body cavities are separated by a flap of tissue. When the pressure of the fluid in the thorax rises, hemolymph is forced into the wing veins. When the flow is reversed, hemolymph leaves the wings; this is aided by the accessory pulsatile organs, mini-hearts that lie at the wing bases.

▶ CIRCULATORY SYSTEM

Nutrients are carried in the hawkmoth's body fluid, or hemolymph, which is pumped around the body cavity by the dorsal vessel. Oxygen is carried to the hawkmoth's cells through a series of tiny tubes called trachea and tracheoles.

The tracheal system

In both caterpillars and adult moths, pores called spiracles allow air to pass through the exoskeleton. They open into a complex, branching network of pressure-resistant air tubes. These are called tracheae. Oxygen diffuses down these tubes, then along even smaller tubes called tracheoles, before eventually reaching the tissues.

The moth's spiracles are closable and may have sievelike structures. These features help reduce water loss and prevent dust from entering the tracheal system.

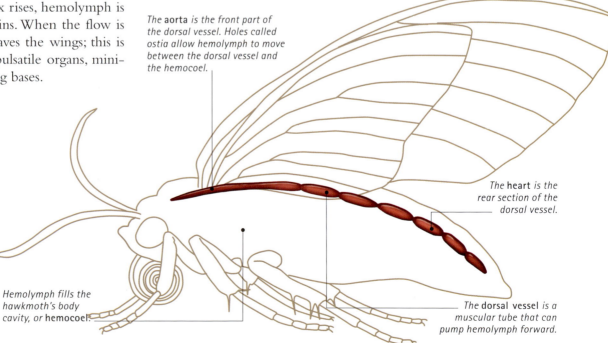

*The **aorta** is the front part of the dorsal vessel. Holes called ostia allow hemolymph to move between the dorsal vessel and the hemocoel.*

*The **heart** is the rear section of the dorsal vessel.*

*Hemolymph fills the hawkmoth's body cavity, or **hemocoel**.*

*The **dorsal vessel** is a muscular tube that can pump hemolymph forward.*

Reproductive system

CONNECTIONS

COMPARE the reusable penis of a hawkmoth with that of a *HONEYBEE,* which snaps off in the female when mating is complete.

COMPARE the life cycle of a hawkmoth with that of a *DRAGON-FLY,* in which only incomplete metamorphosis occurs.

Like other insects, all lepidopterans are either male or female—they are never hermaphrodites (both sexes at once). Males produce sex cells in an organ called a testis. Most animals have a pair of testes, but in lepidopterans there is usually just one. Tubes lead from the testis to the penis, which is inserted into the female during mating. Female lepidopterans have two ovaries. They produce eggs, each of which contains an unfertilized egg cell and a food supply, the yolk. The yolk is produced by the female's fat body (a storage organ in the hemocoel) and is transferred to the ovaries by the hemolymph.

During mating, a male transfers sperm to the female in a packet called a spermatophore. Males have claspers near the end of the abdomen, which they use to hold the female until mating is over. Some hawkmoth pairs remain coupled for several hours during mating. The female stores the sperm inside an organ called the spermatheca. She may mate with several males and therefore possess several spermatophores in her spermatheca.

Female lepidopterans are unusual among insects because most have two openings to their reproductive system. These are on the underside of the body, near the end of the abdomen. One opening is used for the introduction of sperm; the other is the exit through which the insect lays its eggs. As eggs move from the ovary toward this exit, the female releases some sperm from the spermatheca, which fertilizes them. Female lepidopterans attach their eggs to suitable food plants with the aid of secretions from glue-producing glands.

Growing and changing

The first meal of a fresh-hatched caterpillar is often its eggshell. Caterpillars grow quickly, but the exoskeleton does not stretch, so it must be molted at intervals to allow growth. Most caterpillars have five growth stages, or instars, separated by molts. The caterpillar may eat the skin to reabsorb the nutrients it contains.

Caterpillar hemolymph contains a hormone called juvenile hormone (or JH). It tells the

---

COMPARATIVE ANATOMY

Moth and butterfly pupae

Most moths pupate underground, and many also construct a protective cocoon out of silk from their silk glands. The silk used in clothes comes from the cocoons of Chinese silkworm moth pupae. Butterfly pupae do not spin cocoons; they attach themselves to plants above ground before pupating. Butterfly pupae are often called chrysalises. The word comes from the Greek for "golden," because in some species they have a golden sheen. More commonly, however, chrysalises are shaped and colored in such a way that they mimic dead leaves. They either hang upside down from a stalk at their hind end, or form right-way up, held against a surface by a loop of silk like a safety harness.

Pupa
The pupating hawkmoth breathes through a hole in the pupal wall but does not eat, move, or excrete until it has completely transformed into an adult moth.

EGGS
The female hawkmoth lays the eggs on a leaf and glues them to the surface using secretions produced by glands.

CATERPILLARS
The eggs hatch into tiny caterpillars.

ADULT MOTH
The adult moth breaks free of the pupa and pumps hemolymph into its wings.

CATERPILLAR INSTARS
The caterpillar grows rapidly and regularly sheds its skin to grow. Each stage of growth is called an instar. Caterpillars are a pest on some farmed crops such as tobacco and potatoes.

PUPA
After the final instar the caterpillar turns into a pupa and undergoes a complete metamorphosis.

COMPARATIVE ANATOMY

Why pupate?

Insects such as termites, grasshoppers, and dragonflies do not go through pupation. They have a much more gradual change between young and adult called incomplete metamorphosis. Insects of just a few orders undergo complete metamorphosis. However, such orders include all the most successful and species-rich groups, such as the beetles, flies, and lepidopterans. Biologists think that complete metamorphosis evolved just once, leading to a spectacular radiation (expansion in variety) of these groups. What advantage does complete metamorphosis give these insects? Adults and young can live in different places and on different foods, owing to their radically different body plans, so there is no competition for resources between the life stages.

▲ **MOTH LIFE CYCLE**
Like butterflies, moths have a complex life cycle, which includes a complete change in appearance called metamorphosis.

body to remain in juvenile form. Eventually the level of JH falls. The caterpillar then molts into an inactive stage called a pupa. Many insects overwinter as pupae. This stage also sees the massive change in body structure between young and adult. Some structures such as the wings start forming inside the caterpillar, hidden out of view. When the caterpillar becomes a pupa, the half-formed wings fold outward. They are then firmly stuck down on the pupa's surface until they complete their development.

Other adult structures such as the compound eyes grow from tiny groups of cells that become activated in the pupa. At the same time, the pupa breaks down and recycles many of the old caterpillar tissues. When the metamorphosis (change) is complete, the adult emerges through the pupal skin. The wings are folded and soft, and the adult moth is therefore very vulnerable to predators. However, the adult swiftly expands its wings by pumping hemolymph into the wing veins.

RICHARD BEATTY

FURTHER READING AND RESEARCH

Himmelman, John. 2002. *Discovering Moths.* Down East Books: Camden, ME.

Kitching, Ian J., and Jean-Marie Cadiou. 2000. *Hawkmoths of the World.* Cornell University Press: Ithaca, NY.

Scott, James, A. 1992. *The Butterflies of North America.* Stanford University Press: Stanford, CA.

Hippopotamus

ORDER: Artiodactyla FAMILY: Hippopotamidae
GENERA: *Hippopotamus* and *Hexaprotodon*

There are two species of hippopotamuses, or hippos. Both species live in sub-Saharan Africa. The common hippo spends the day resting in rivers or pools, emerging at night to feed on land. The pygmy hippo lives in shaded forests.

Anatomy and taxonomy

The word hippopotamus means "river-horse," though these large, barrel-bodied, hoofed mammals are more closely related to pigs than to horses.

● **Animals** All animals are multicellular and rely on other organisms for food. They differ from other multicellular life-forms in their ability to move around (generally using muscles) and their rapid responses to stimuli.

● **Chordates** At some stage in its life, a chordate has a stiff, dorsal (back) supporting rod called a notochord.

● **Vertebrates** The notochord of vertebrates develops into a backbone (also called spine or vertebral column), which is made up of connected bones called vertebrae. The vertebrate muscular system that moves the head, trunk, and limbs consists primarily of muscles that are bilaterally symmetrical around the skeletal axis. In other words, the muscles on one side of the backbone are the mirror image of those on the other side.

● **Mammals** Members of the class Mammalia share several features. They are endothermic animals (warm-blooded, with the warmth generated internally), with a body usually covered with hair or fur and a lower jaw formed by a single bone that hinges directly with the skull. Female mammals nourish their newborn on milk from mammary glands. Mammalian red blood cells differ from those of other vertebrates in that they do not contain a nucleus.

● **Even-toed ungulates** Most of the 220 or so species of hoofed mammals are artiodactyls, or even-toed ungulates. They are so named because each foot bears two or four hoofed toes. Sheep, cattle, deer, camels, pigs, and hippos are all even-toed ungulates. These animals have a complex and highly efficient digestive system that enables them to make efficient use of a mainly vegetarian diet. The gut of an even-toed ungulate contains symbiotic bacteria or other single-celled organisms. These organisms help break down cellulose, the chemical that gives plant cell walls their

▼ This family tree shows the two living species of hippos and their relationships with the other even-toed ungulates.

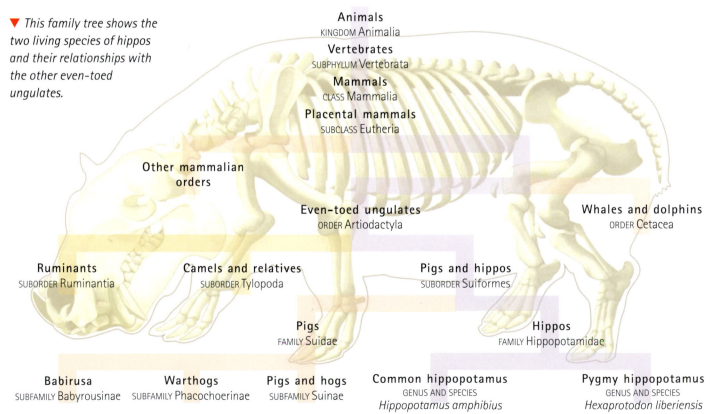

Animals
KINGDOM Animalia

Vertebrates
SUBPHYLUM Vertebrata

Mammals
CLASS Mammalia

Placental mammals
SUBCLASS Eutheria

Other mammalian orders

Even-toed ungulates
ORDER Artiodactyla

Whales and dolphins
ORDER Cetacea

Ruminants
SUBORDER Ruminantia

Camels and relatives
SUBORDER Tylopoda

Pigs and hippos
SUBORDER Suiformes

Pigs
FAMILY Suidae

Hippos
FAMILY Hippopotamidae

Babirusa
SUBFAMILY Babyrousinae

Warthogs
SUBFAMILY Phacochoerinae

Pigs and hogs
SUBFAMILY Suinae

Common hippopotamus
GENUS AND SPECIES
Hippopotamus amphibius

Pygmy hippopotamus
GENUS AND SPECIES
Hexaprotodon liberiensis

strength, into useful sugars. Even-toed ungulates also possess effective antipredation strategies: many are fast runners, and some have horns, hard hooves, or tusks.

▲ *Since its eyes, ears, and nose are situated high on its head, a common hippopotamus is able to survey the environment even when most of its body is submerged.*

● **Piglike artiodactyls** All members of this group have a large head, a barrel-shaped body, a complex chambered stomach, and low-crowned cusped teeth.

● **Pigs** Pigs are medium to large mammals with a large, tapering head, slim legs, and small feet. Their skin is covered with a sparse scattering of coarse, bristly hair. Only the middle two toes on each foot touch the ground.

● **Peccaries** The three species of peccaries live in South and Central America. They have relatively slender legs and

small feet, with two functional hoofed toes on each foot. The head is very large relative to body size, and the whole body is covered with a coat of coarse hair.

● **Hippopotamuses** Members of this family are large to very large, with a robust body, short stocky legs, a large head with a broad snout, a large mouth, and a short, narrow tail. The thick skin is almost completely hairless, and usually appears oily. Each foot has four functional hoofed toes. There are only two species: the common hippopotamus and the pygmy hippopotamus.

EXTERNAL ANATOMY A large, virtually hairless, roughly cylindrical body on short, stout legs. The head is very large with a wide mouth and large teeth. *See pages 508–511.*

SKELETAL SYSTEM The skeleton is stout and sturdy. It has a particularly large, heavy skull and a massively reinforced lower jaw. *See pages 512–515.*

MUSCULAR SYSTEM The heavy skeleton is matched by powerful muscles, especially around the neck and shoulders. Hippos are strong rather than fast. They are able to travel long distances on land and move around easily in water. *See pages 516–517.*

NERVOUS SYSTEM The brain is small compared with that of other artiodactyls, but the senses of sight, hearing, and smell are well developed. *See pages 518–519.*

CIRCULATORY AND RESPIRATORY SYSTEMS Respiration and circulation are highly efficient, with large lungs, a slow breathing rate, and a slow heart rate when at rest. These features help hippos save energy and allow them to stay underwater for several minutes. *See pages 520–521.*

DIGESTIVE AND EXCRETORY SYSTEMS Hippos are vegetarians. They do not chew the cud, but their food passes though a complex stomach and is digested with the help of single-celled organisms that live in the gut. *See pages 522–523.*

REPRODUCTIVE SYSTEM Hippopotamus sex organs are typical of artiodactyls. Males have internal testes and a retractable penis; females have a Y-shaped uterus and paired mammary glands. *See pages 524–525.*

External anatomy

CONNECTIONS

COMPARE the positioning of nostrils, eyes, and ears on the hippopotamus head with that of another semi-aquatic mammal, the *OTTER*.

COMPARE the greasy-looking skin of a hippo with the thick hide of other large, nonhairy mammals such as *ELEPHANTS* and *RHINOCEROSES*.

The common hippopotamus is the largest species of even-toed ungulate. Males are slightly larger and substantially heavier than females. Average weights are just under 1.65 tons (1.5 metric tons) for females and between 1.65 and 2 tons (1.5–1.8 metric tons) for males, though exceptionally large individuals may exceed 3.3 tons (3 metric tons).

The common hippopotamus in particular has a disproportionately large head. The head is supported by a short, very thick neck, which rises from a massive chest and muscular shoulders. The gargantuan mouth, which opens to almost 150 degrees, contains some of the largest teeth of any mammal: the two massive lower canines.

Skin deep

Hippopotamuses are virtually bald, with just the sparsest scattering of body hair in both species. The skin of the common hippo is

▶ **Common hippopotamus**
This species is the larger of the two hippos. It spends the hours of daylight in rivers and lakes with firm, sloping borders. At night it grazes on land.

*The hippo is able to tuck its **earflaps** down when it goes underwater. This keeps water from entering the ears.*

*The **eyes** are positioned high on the head. The hippo can see its surroundings even when mostly submerged.*

The nostrils protrude, so they are above water when the rest of the animal is submerged.

*The **mouth** is able to open 150 degrees.*

usually dark brown or purple and is sometimes mottled pinkish brown around the belly. Hippopotamus skin can be up to 2.5 inches (6 cm) thick on the back, flanks, and rump. Even on the belly and the insides of the legs the skin is at least 0.5 inch (1.25 cm) thick. For this reason taxonomists (scientists who classify living things) once placed hippos in a group called the pachyderms, alongside other thick-skinned mammals such as elephants and rhinoceroses. Hippopotamus leather is used for making headdresses and African *sjambok* whips.

Most of the thickness of the skin is made up of the dermis layer, which contains a tightly

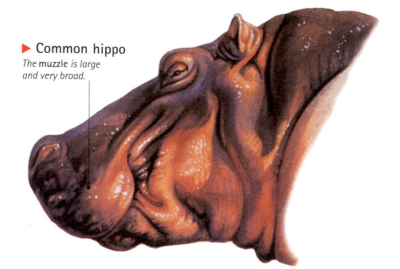

▶ **Common hippo**
The muzzle is large and very broad.

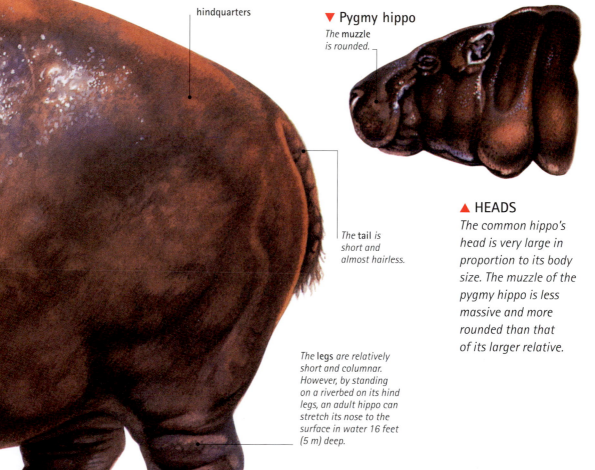

hindquarters

▼ **Pygmy hippo**
The muzzle is rounded.

The tail is short and almost hairless.

▲ **HEADS**
The common hippo's head is very large in proportion to its body size. The muzzle of the pygmy hippo is less massive and more rounded than that of its larger relative.

The legs are relatively short and columnar. However, by standing on a riverbed on its hind legs, an adult hippo can stretch its nose to the surface in water 16 feet (5 m) deep.

59–65 inches (150–165 cm)

118–197 inches (300–500 cm)

The barrel-shaped body is covered with tough hide. This is thickest around the hindquarters.

The four toes are connected by small webs of skin.

packed mat of tough collagen fibers and a thick layer of fat. The outermost layer of hippo skin, called the epidermis, is very thin. This structure has important implications for hippo behavior. Unlike the tough hide of most other hoofed animals, hippopotamus skin is far from waterproof. Hippos lose water at a great rate—several times faster even than a sweating human despite having no sebaceous (sweat) glands and thus being unable to perspire. They lose water purely by evaporation through the skin. If allowed to dry out, a hippo's skin soon becomes cracked and sore. The common hippo avoids dehydration (losing too much water) and stays cool by spending the day wallowing in rivers and lakes. Hippos generally emerge only at night or when it rains.

The pygmy hippopotamus gets around the problem of water loss by spending most of its life in dense forest, avoiding the direct sunlight that could dry and damage its skin.

Another important factor in maintaining healthy skin is the attention of certain cleaner fish, mostly carps such as those in the genus

COMPARATIVE ANATOMY

The pygmy hippopotamus

The smaller surviving member of the family Hippopotamidae is the pygmy hippopotamus. It is considerably smaller than the common hippo, with the largest individuals weighing about 600 pounds (270 kg). Pygmy hippos are more piglike than common hippos, with a slimmer body, a smaller head, and a smooth, oily, dark brown to olive-gray skin. The name of the pygmy hippo's genus, *Hexaprotodon*, means "six first teeth." This is misleading, since pygmy hippos have only one pair of incisors in the lower jaw. However, the fossil relatives from which the genus was first described had three pairs of incisors, making six front teeth in each jaw. European scientists did not know the pygmy hippopotamus existed until the 19th century.

Common hippo   Pygmy hippo

COMPARATIVE ANATOMY

Unusual hooves

Hippopotamuses are unusual in that each foot has four hoofed toes, all of which make contact with the ground. Their closest cousins, the pigs, also have four toes, but only two are functional and make contact with the ground. Hippo footprints are therefore unique, and at up to 11 inches (28 cm) long cannot be easily mistaken for those of any other animal.

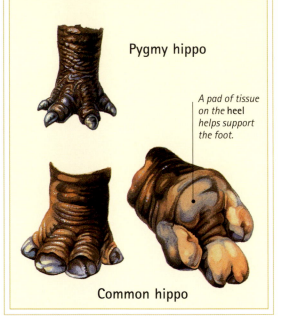

Pygmy hippo

A pad of tissue on the heel helps support the foot.

Common hippo

Labeo, which nibble dirt and debris from the hippos' bodies as they rest in the water. Grooming birds such as oxpeckers and egrets perform a similar job; they perch on the backs of resting hippos and pick off parasites.

Adaptations to water

Common hippopotamuses have a number of features that help them live in water. Most of the animals' time is spent in relatively shallow water, standing on the bed of a river or lake. A common hippo's feet are large, with partially webbed toes that can spread a little. This helps disperse the animal's weight over a large area so it does not sink deeply into the mud at the water's edge. Once a hippo is submerged, much of its great weight is supported by the water, allowing it to move easily and with surprising grace for such a big animal.

Sweating blood?

Hippopotamuses secrete a pink, greasy substance from pores in their skin. These pores are large enough to see with the naked eye, and they are spaced at a density of around 4 or 5 per square inch (1 per cm²). Each pore connects to a small gland. People once thought hippos sweated blood, but they were wrong on two counts: hippos have no sweat glands, and the substance is not blood but a kind of salty mucus. This dries to form a shiny lacquer over the hippo's skin. Biologists now think this mysterious secretion serves as a kind of sunscreen, helping keep the hippo's hairless skin from getting sunburned.

Perhaps surprisingly, hippos are poor swimmers. They do swim, using a kind of dog-paddle, but they prefer not to get out of their depth. Even in water that is deep enough for swimming, they usually progress in a series of bounds, pushing off from the bottom and breaking the surface with each leap. The legs and feet are easily strong enough to bear the animal's considerable weight on land. They are located under the four corners of the body like stout pillars. All four toes on each foot touch the ground when the hippo is walking.

In ideal circumstances, hippos will opt to rest in water less than 5 feet (1.5 m) deep. Then they can stand or lie on the bottom with the top of the head and the snout resting at the surface. It is no coincidence that a hippo's ears, eyes, and nostrils are all located on the top of the head and snout. The arrangement of the sensory organs allows the animal to breathe easily, and to watch and listen to the sights and sounds around it while remaining almost completely submerged.

▲ *These hippos are basking in the sunshine. If a hippo spends too much time exposed to the sun's rays, its skin will dry and crack. Thus, during the day hippos mostly wallow in water.*

Changing shape over time

Once, there were several other species of hippos, and they inhabited a far more extensive range. One million years ago, there were at least eight species of hippos in Africa and several in Asia. Most species have disappeared over the last few thousand years. Those were mainly species that lived on islands. Island animals also adopt forms very different from their ancestors relatively quickly; new species are able to evolve in a relatively short period. With few predators on islands, small mammals tend to become much larger. The islands in the Mediterranean Sea were once home to giant shrews, hedgehogs, and dormice. By contrast, large mammal species on islands tend to shrink dramatically over time in response to a much reduced availability of food and territory ranges. There were dwarf hippos and elephants on islands in the Mediterranean; the elephants were less than a fifth of the size of their mainland relatives. These animals became extinct soon after people colonized the islands, around 10,000 years ago. Around 1,000 years ago three species of dwarf hippos on the Indian Ocean island of Madagascar became extinct.

Skeletal system

Both species of hippos have a rotund body shape and a robust skeleton. As with other large mammals, the backbone is made up of large interlinked vertebrae with tall vertical projections called thoracic spines. These extensions of the dorsal skeleton allow for the attachment of large muscles. The hippo's skeleton has many adaptations for bearing heavy loads, including thick limb bones, vertical shoulder blades, and thick ankle and wrist bones. In engineering terms this type of structure is called "mediportal." These features are not as pronounced as in elephants, which have a "graviportal" structure. The hippo's

▶ **Common hippopotamus**
The major bones of the head, the cranium and the mandible, are massive. The head is supported by the strong cervical vertebrae in the neck. The bones of the front and hind legs are also thick and strong.

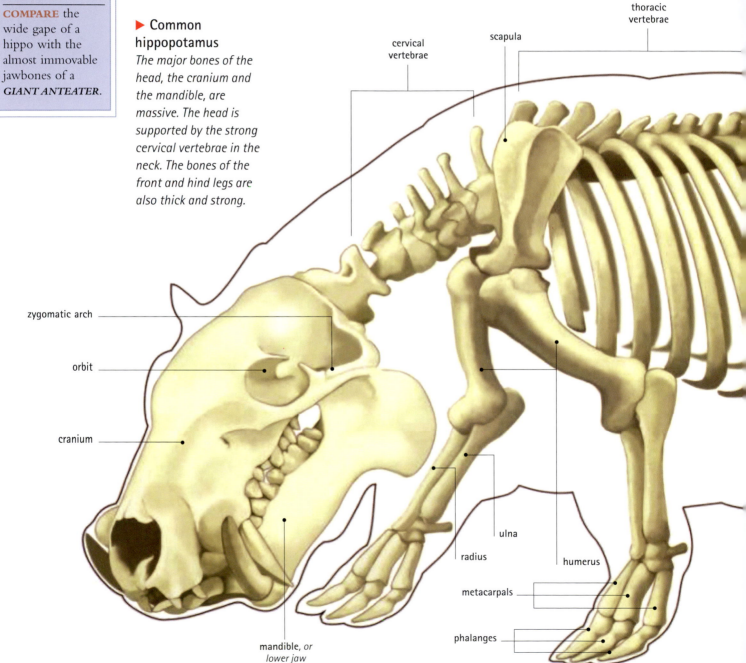

cervical vertebrae

scapula

thoracic vertebrae

zygomatic arch

orbit

cranium

ulna

radius

humerus

metacarpals

phalanges

mandible, *or lower jaw*

512

skeleton is the living equivalent of a simple single-girder bridge, with the vital organs slung below. The rib cage, with 13 pairs of sturdy ribs, is large to accommodate the massive lungs.

Limbs and feet

Common hippopotamus limb bones are short and stout, but perhaps not as robust as might be expected for an animal of such great size. Hippos spend much of their time in water where less stress is placed on the legs. One of the features that distinguish hippos from their closest relatives the pigs, and from other hoofed animals, is the structure of their feet. Their hooves are relatively small and soft, a bit like

Heavy bones

Hippopotamus bones are unusually heavy. They contain a greater density of mineral matter than the bones of most other mammals. They act as ballast, helping offset the buoyancy of the rest of the body in water. This allows the hippo to walk comfortably on the bed of a river or lake. Animals with lighter skeletons, such as humans, struggle to walk in deep water because their bodies have a natural inclination to float. Hippos share the useful adaptation of bone density with the manatees and dugongs, aquatic mammals that graze on sea grass and other bottom-growing water weeds.

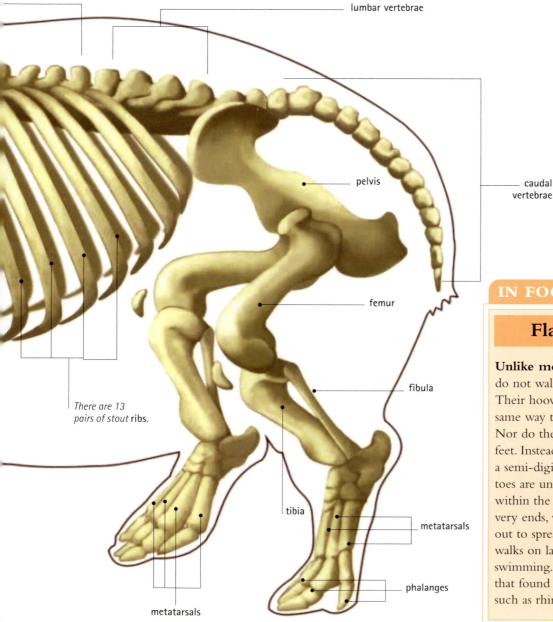

lumbar vertebrae

pelvis

caudal vertebrae

femur

fibula

There are 13 pairs of stout ribs.

tibia

metatarsals

phalanges

metatarsals

IN FOCUS

Flat feet or tiptoes?

Unlike most other artiodactyls, hippos do not walk on the very tips of their toes. Their hooves are not weight-bearing in the same way that those of cattle or deer are. Nor do they walk on the flat sole of their feet. Instead they have evolved what is called a semi-digitigrade stance; the bones of the toes are unfused but are bound together within the tissue of the foot, except for the very ends, which are separate. The toes splay out to spread the animal's weight when it walks on land, and close together for swimming. This arrangement is similar to that found in certain odd-toed ungulates such as rhinoceroses.

large toenails. They are not part of the skeleton, since they are made not of bone, but of a protein called keratin. This is the same protein from which the hair and nails of all other mammals are made.

The skull

The massive skull of the common hippopotamus bears a tall ridge called a sagittal crest, which runs along the top of the skull. The skull has large cheek arches to accommodate some of the extra-large muscles that control the jaw. The brow ridges are high, reflecting the position of the eyes on top of the head, and the braincase is small.

The lower jaw is much wider than the upper part of the skull. This is necessary to accommodate the giant, tusklike lower canine teeth. These are much too long to fit inside the hippo's mouth.

When a hippo opens its mouth wide, the upper jaw (and with it the whole skull) lifts up and back. Compare this with the movement of the human jaw, in which a wide gape involves mostly a downward movement of the lower jaw. The reason hippos do things differently is strength. A loose, mobile lower jawbone such as that of a human is not good at taking impacts—it can be dislocated or even broken by a heavy blow. The heavy lower jaw of a hippo has powerful muscular attachments to

the skull and neck and is incredibly strong. Not only can it withstand powerful impacts when fighting other hippos, but it can also be used to exert massive, prolonged pressure as rivals wrestle with their teeth locked together.

Because of the awesome power of hippo jaws, most other animals give them a very wide berth. There are numerous accounts of hippos taking bites of out of wooden fishing boats, and even a report of an adult common hippo biting a 10-foot (3-m) crocodile in two.

Mighty teeth

In some regards, hippos have a typical artiodactyl dentition. There are 38 (sometimes 42) teeth in total, with the front teeth (the incisors and canines) separated from the cheek teeth (the molars and premolars) by a wide gap called the diastema.

The common hippo has two pairs of incisors in each jaw. Those in the upper jaw are short and rounded in cross section and point downward. In the lower jaw they are more triangular in cross section and project forward 6 or 7 inches (17 cm) from the gum; the entire tooth, including the root embedded in the

▼ SKULL
Common hippopotamus
The lower jaw is large relative to the upper jaw, providing plenty of surface area for the attachment of muscles. Note how the nasal cavities and eye sockets are both high on the skull. This position allows the hippo to breathe, and see and smell predators, even when mostly immersed in water.

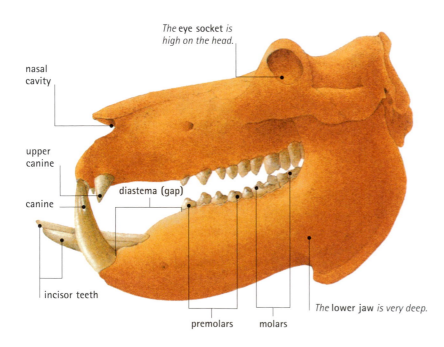

The eye socket is high on the head.

nasal cavity

upper canine

canine

diastema (gap)

incisor teeth

premolars

molars

The lower jaw is very deep.

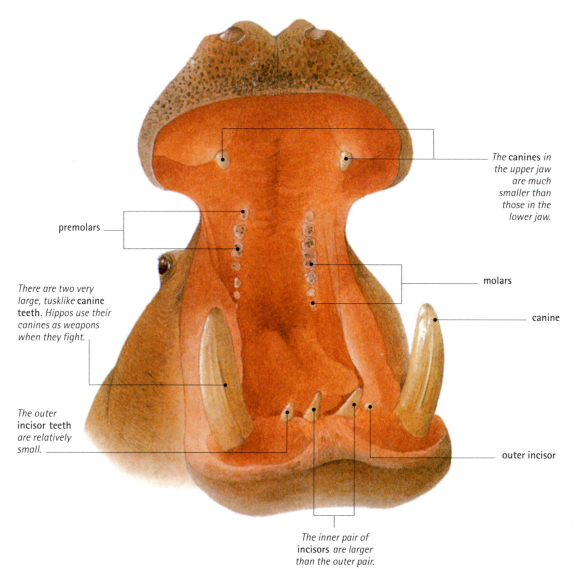

◀ TEETH
Common
hippopotamus
The massive lower
canines may project
up to 12 inches (30 cm)
from the gum.

*The **canines** in*
the upper jaw
are much
smaller than
those in the
lower jaw.

premolars

molars

There are two very
*large, tusklike **canine***
***teeth**. Hippos use their*
canines as weapons
when they fight.

canine

The outer
incisor teeth
are relatively
small.

outer incisor

The inner pair of
incisors *are larger*
than the outer pair.

gum, is around 12 to 14 inches (30–40 cm) long. The forward-projecting arrangement of the lower incisors allows rival animals to lock jaws when sparring, in much the same way that male deer lock antlers.

Even more conspicuous than the incisors, especially on males, are the tusklike lower canine teeth. In a fully grown male, these protrude up to 12 inches (30 cm) from the gum and have roots that reach up to 16 inches (40 cm) into the lower jaw. These very deep roots anchor the tooth firmly and allow it to withstand huge forces. The canines are triangular in cross section and very sharp. The upper and lower canines slide together as neatly as scissor blades as the animal opens and closes its mouth; the continual strafing maintains a razorlike edge. Male and female

hippos use their mighty canines for fighting; the wounds they can inflict are so terrible that often just the sight of the lethal teeth is enough to intimidate a rival into backing down. Display is therefore an important part of social interaction. Hippos frequently adopt a wide-mouthed gape, so that other hippos can see their weapons. Older males are often seen with broken canines as a result of brutal battles.

The cheek teeth, of which there are four pairs of premolars and three pairs of molars, are used for feeding. Tough vegetable matter is ground between the low-crowned surfaces of the teeth. The teeth grow continuously to compensate for the constant wear and tear at the grinding surface. Each millstone-like grinding tooth is roughly square, usually with a raised cusp in each quarter.

Muscular system

CONNECTIONS

COMPARE the huge muscles of a hippo's neck with those of a *GIRAFFE*. The giraffe's neck muscles are long, but they do not have to support the great weight of massive jaws.

Hippos are immensely strong. When they are immersed in a river or lake, their body weight is largely supported by the water. However, unlike more fully aquatic large mammals, such as seals and whales, hippos also need to move around extensively on land. This makes strong legs essential.

Head and neck muscles

Supporting the weight of the huge head, especially out of water, also requires great strength. The neck and shoulders are very muscular. The temporalis and masseter muscles, which control the movement of the massive lower jaw, are huge. They can be seen bulging at the back of the neck as the animal chews or opens its mouth. The heavy jowls conceal the muscles that attach the lower jaw to the base of the skull and the neck.

Smaller jaw muscles, called the pterygoids, provide the side-to-side and rotational movement required to grind plant material

between the back teeth. Another set of muscles important in feeding make up the tongue. The tongue is mobile, especially in young animals that are still suckling. They use the tongue to grasp the mother's teat when nursing.

The ears and nostrils have small muscular valves that allow them to close when the hippo goes underwater. The external ears also have muscles that allow them to rotate to follow sounds, and to be flicked back and forth rapidly to clear them of water or mud.

Walking muscles

When a hippo walks, it is the upper limb muscles that do most of the work. The shoulder and hip joints are more mobile than the elbows and knees or the wrists and ankles. The muscles of the lower limb are used to lift the feet, but the power to swing the legs and propel the animal comes from higher up. The massive shoulder and hip muscles of a hippo extend right around the body, and attach not

▶ **SUPERFICIAL MUSCLES**
Common hippopotamus

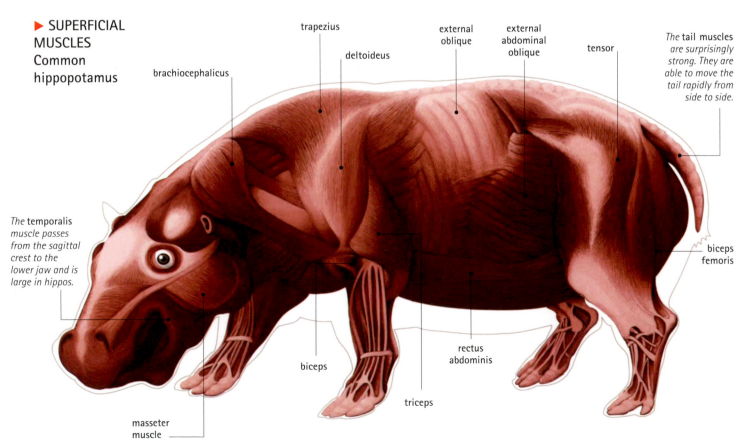

trapezius

deltoideus

brachiocephalicus

external oblique

external abdominal oblique

tensor

The **tail muscles** *are surprisingly strong. They are able to move the tail rapidly from side to side.*

The **temporalis** *muscle passes from the sagittal crest to the lower jaw and is large in hippos.*

biceps femoris

biceps

triceps

rectus abdominis

masseter muscle

only to the shoulder and pelvis bones but also to the vertebral column. Common hippos are able to gallop at speeds of up to 25 miles per hour (40 km/h) over short distances, easily outstripping the fastest human athletes. Since hippos are almost always seen wallowing in water by day, it is easy to forget that they spend most of the night moving around on land. They are well adapted for walking. Their stout legs are robust and powered by strong muscles. Hippos may wander up to 21 miles (33 km) in a night in search of good grazing.

The tail, although small, is very muscular. It is flattish, and it can be whipped back and forth almost faster than the eye can see. This motion is used to drive biting insects away and to spread feces during territorial marking.

▼ *The hippo jaw can be opened very wide, much more so than is needed for feeding. The reason for this is so a hippo can show off its lower canines, which are formidable weapons in territorial battles.*

IN FOCUS

Natural born healers

Being aggressive animals, hippos often sustain serious flesh wounds during fights. Given the amount of time hippos spend in dirty water, one might expect that these wounds would often become infected. Surprisingly, that is rarely the case. Indeed, hippos seem to have amazing healing abilities. It may be that the sticky skin secretion that helps condition the hippo's skin also has antiseptic properties that help keep infections at bay. Because wounds are permanently moist, crusty scabs cannot form, so flesh wounds heal with minimal scar tissue produced. The same principle has led to the development of hydrocolloid dressings for first aid in humans. The dressings promote rapid healing by keeping wounds moist.

Nervous system

A hippo's nervous system is able to receive and analyze external stimuli—sights, sounds, and smells, for example—and coordinate the animal's body to react appropriately. The core of a hippo's nervous system is its brain and spinal cord. Together, these features make up the central nervous system (CNS). Other nerves, connecting every part of the animal's body with the CNS, make up the peripheral nervous system (PNS).

The brain of a common hippopotamus is small relative to the size of its body. The two cerebral hemispheres, which together form the cerebrum, are small. The cerebrum probably plays an important role in the process of learning and in short-term and long-term memory. The cerebrum does not cover the cerebellum, which is the section of the brain associated with the coordination of muscle activity and balance. However, the sections of the brain associated with the sense of smell are large; in social interactions between hippos, smell is probably the most important sense.

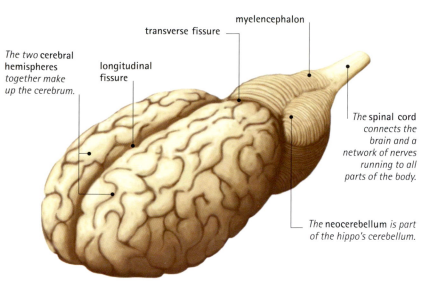

The two cerebral hemispheres *together make up the cerebrum.*

transverse fissure

myelencephalon

longitudinal fissure

The spinal cord *connects the brain and a network of nerves running to all parts of the body.*

The neocerebellum *is part of the hippo's cerebellum.*

Aggression in defense

Hippos are not considered to be intelligent because most of their behavior is instinctive or reactive. They are easily annoyed. In all but territorial adult males and females with young, the first reaction to a perceived threat is to move away. However, if the escape route is

▲ BRAIN
Common hippo
The cerebrum is responsible for learning and memory. The cerebellum coordinates movement and balance.

► **Common hippopotamus**

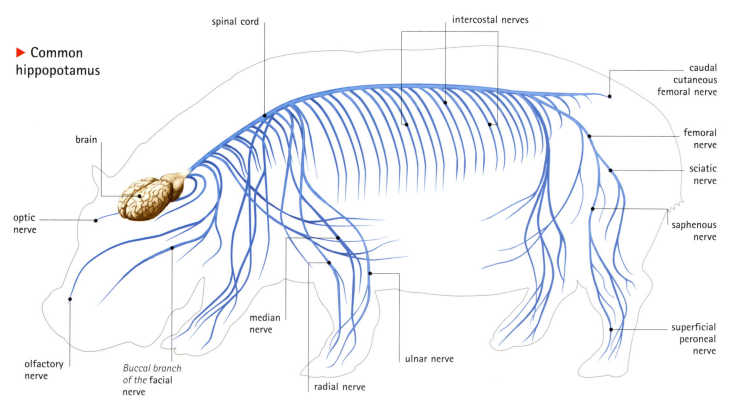

spinal cord

intercostal nerves

caudal cutaneous femoral nerve

brain

femoral nerve

sciatic nerve

optic nerve

saphenous nerve

median nerve

superficial peroneal nerve

olfactory nerve

Buccal branch of the facial nerve

ulnar nerve

radial nerve

Long-distance calls

Hippos can produce and detect infrasound, very low-frequency sound that humans cannot hear. Infrasound is ideal for communication over very long distances—up to 20 miles (32 km) in water. Although sounds produced underwater cannot easily cross to the air, and vice versa, that is not a problem for hippos, which produce sound from different parts of the body. A hippo can bellow through its nostrils to emit infrasound, and sounds that humans can hear, into the air. To produce sounds underwater, the hippo squeezes air in the throat, and sounds are channeled through a blob of blubber under the jaw. When a hippo is in water and its ear-flaps are closed, it can still detect infrasound vibrations through its lower jaw. A hippo floating at the surface is able to judge the distance of a source of infrasound, such as a bellowing hippo. This is possible because sound travels faster in water than in air. The greater the interval between the arrival of infrasound at the jaws and at the ears, the farther away the other hippo is.

blocked or the threat continues to approach, hippos use attack as the best form of defense and charge at the enemy. Territorial males are more likely to attack a perceived threat without first being provoked.

Hippo senses

The common hippo has eyes that protrude from the top of the head and bulge out to the sides, so the animal has a very wide field of vision. This arrangement is not very good for judging distance or seeing in three dimensions; these skills require forward-facing eyes like those of a primate or a big cat. However, the hippo arrangement is excellent for all-round awareness and for picking up small movements. Common hippos can see almost as well at night as they can by day. Optic nerves connect the eyes and brain.

Hippos have a good sense of hearing. Their small, highly mobile earflaps can be moved around to track quiet, distant sounds. Acoustic nerves connect the ears and brain. Hippos use numerous different sounds to communicate. Most of these are produced by rapid expulsion of air from the lungs, resulting in various grunts, huffs, and roars and a characteristic sound known as "wheeze-honking," which signifies excitement and is commonly used during courtship.

Detecting pheromones

Hippos also have a "sixth sense." Many male ungulates, including hippos, curl their upper lip when breathing the scent of a female. This behavior is called the Flehman response. Zoologists believe it enables the male to detect minute quantities of pheromones, chemicals that tell the males how close a female is to being sexually receptive. This sense is related to smell and taste but does not use the nose or taste buds. Instead, the chemical clues are picked up by the vomeronasal, or Jacobson's, organ, located in the roof of the mouth.

▲ *Adult male hippos are able to detect the scent of a female that is ready to breed. At this time, the males may fight vigorously for the right to mate.*

Circulatory and respiratory systems

CONNECTIONS

COMPARE the involuntary breathing of a hippopotamus with the voluntary breathing of a *DOLPHIN* or *GRAY WHALE*.

COMPARE the length of time a hippo can spend underwater with the dive period of a *SEAL*.

As is to be expected for a very large animal, hippos have two large lungs. However, hippos also have a relatively low metabolic rate. As a result, pound for pound they use less energy and require less oxygen than, for example, a smaller mammal such as a jackrabbit or a small monkey. Even when small mammals are resting, they breathe rapidly and their heart beats fast to keep up with the demand for resources. A hippo's body runs so efficiently that, weight for weight, its metabolic rate is comparable to that of a cold-blooded animal like a medium-sized lizard.

Above water, hippos breathe in and out (inhale and exhale) six or seven times a minute. Scientists have recently discovered that large, slow-breathing mammals such as elephants and hippos often hold their breath for short periods when resting. So instead of happening at regular intervals, those six or seven breaths per minute may come in fairly quick succession, followed by a long pause and a large exhalation. Long, slow, continuous

muscular contractions require more effort than short intermittent ones. So for a hippo, holding its breath occasionally is a convenient way of saving precious energy.

Easy breathing
Breathing is an involuntary reflex for hippos, just as it is for humans. An involuntary reflex is an action an animal does not have to think about. The hippo does not have to make a decision before breathing in and out: the process happens automatically. This involuntary regulation of breathing is typical of terrestrial and semiaquatic mammals. In truly aquatic mammals, such as dolphins, breathing is voluntary: the animal must remember to breathe when it comes to the surface.

Diving underwater
Compared with more fully aquatic mammals, hippos are not accomplished divers. They can hold their breath comfortably for about six minutes, a longer time than most humans

▼ **Common hippopotamus**
The organs of the respiratory and circulatory systems are shown, and the most important arteries and veins are depicted.

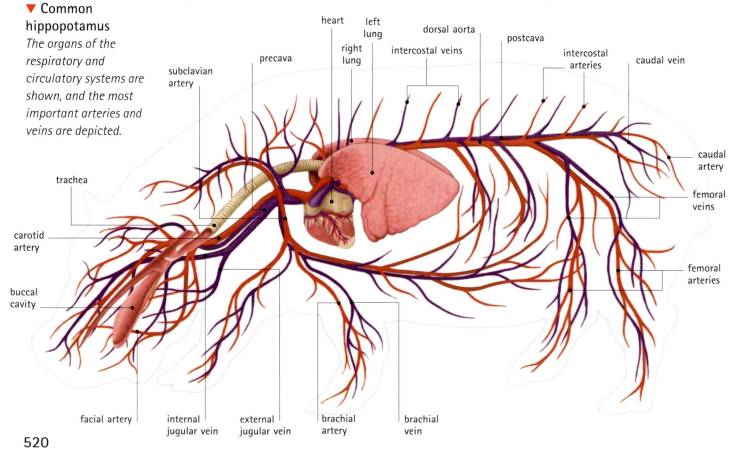

heart · left lung · dorsal aorta · postcava · right lung · intercostal veins · intercostal arteries · caudal vein · precava · subclavian artery · trachea · caudal artery · femoral veins · carotid artery · buccal cavity · femoral arteries · facial artery · internal jugular vein · external jugular vein · brachial artery · brachial vein but not long compared with seals, dolphins, or whales. When diving, a hippo holds its breath voluntarily, but muscles that seal off the nostrils contract as a reflex. This keeps water from entering the lungs.

The heart

The hippo's heart is a powerful muscle that pumps blood around the body. Among many functions, the blood carries oxygen to the cells and disposes of waste carbon dioxide, the product of the animal's metabolism. Like that of other mammals, the hippo's heart consists of four chambers. There are two thin-walled atria and two thick-walled ventricles. During the pulmonary stage of the blood's circulation, the right atrium receives oxygen-deficient blood from the body. Blood then passes into the right ventricle and is pumped to the hippo's lungs, where it picks up a fresh supply of oxygen and loses carbon dioxide. In the systemic circulation, the blood, which is now rich in oxygen, returns to the heart's left atrium. The blood flows into the left ventricle and is then pumped around the body.

A common hippo's heart usually beats about 70 times a minute, a rate similar to that of humans. However, as with many other diving mammals, this rate slows dramatically when the hippo is submerged. Fewer than 20 beats per minute is a normal diving heart rate for a young, healthy hippopotamus.

Keeping cool

For animals that live in cold climates, a massive body is an advantage in keeping warm. Large animals have a relatively small ratio of surface area to volume; in other words, there is less skin per unit of volume than in small animals. With a small ratio, heat is lost more slowly across the skin.

This arrangement is an advantage for a whale in a cold polar sea, but for a large animal in a hot climate, such as a hippopotamus, a slow cooling rate is far from advantageous. To avoid overheating, hippos must find alternative means of losing excess heat. The skin, while prone to sunburn, contains many fine blood vessels or capillaries, which bring warm blood close to the body surface. When hippos are immersed in water, which conducts heat approximately 25 times more effectively than air, this arrangement allows them to rid themselves of large amounts of metabolic heat.

CLOSE-UP

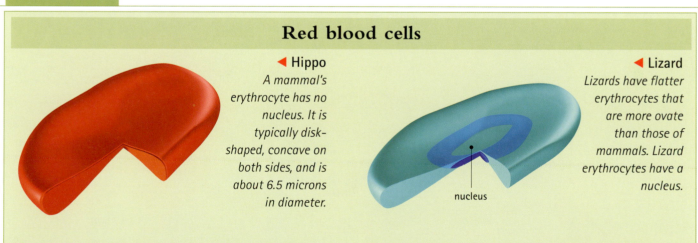

Red blood cells

◀ **Hippo**
A mammal's erythrocyte has no nucleus. It is typically disk-shaped, concave on both sides, and is about 6.5 microns in diameter.

nucleus

◀ **Lizard**
Lizards have flatter erythrocytes that are more ovate than those of mammals. Lizard erythrocytes have a nucleus.

The red blood cells of hippos and other mammals are very different from those of non-mammal vertebrates, such as reptiles, amphibians, and birds. Red blood cells, or erythrocytes, are the most abundant cells in a hippo's blood. They contain the hemoglobin molecule, which is able to hold oxygen and carry it to cells that need it. A hippo's red blood cells—like those of other mammals—are shaped like disks that are concave on both sides. They can change shape to an amazing extent, without breaking, as they squeeze single file through the tiniest capillaries, minute blood vessels through which oxygen, nutrients, and waste products are exchanged throughout the body. The strangest thing about a mammalian red blood cell is that it has no nucleus. The nucleus is extruded from the cell as it matures.

Digestive and excretory systems

Like other even-toed ungulates, hippos are vegetarian, although they also eat carrion (dead animals) sometimes. Hippos graze grass using their muscular lips, which, being straight and up to 20 inches (50 cm) wide, are able to crop a broad swath of grass as closely and efficiently as a lawn mower. Hippos will return to these cropped areas often, preferring the tender new shoots to tough, older stems. A large individual can consume more than 220 pounds (100 kg) of plant material, mostly grass, in a single night. Hippos drink water from the rivers where they rest.

Most vegetable matter, even tender young grass, has a relatively low energy and nutrient content, so hippos must digest their food very thoroughly in order to grow and stay healthy. Most artiodactyls are ruminants. They kick-start the digestive process by chewing cud: they regurgitate food that has already been mixed with juices from the first part of the stomach and chew it again to help speed the digestive process along. Hippos and pigs, however, are not ruminants and do not chew the cud. Instead they rely on features of the stomach to process food effectively.

Stomach complexity

Hippos have arguably the most complex stomach system of all artiodactyls, but they have the simplest intestine. The stomach is huge, with four distinct chambers. The first two form large adjacent pouches called the right diverticulum and left diverticulum. These pouches contain large numbers of single-celled organisms called ciliates; they help break down tough plant compounds, such as cellulose.

From the diverticula, food passes first into the anterior (or cardiac) stomach, then to the posterior (or pyloric) stomach. In effect, the first three chambers are holding vessels, with a combined capacity of up to 105 gallons (400 l), more than 75 percent of the volume of the entire digestive system. Food collected in a

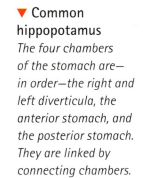

▼ Common hippopotamus
The four chambers of the stomach are—in order—the right and left diverticula, the anterior stomach, and the posterior stomach. They are linked by connecting chambers.

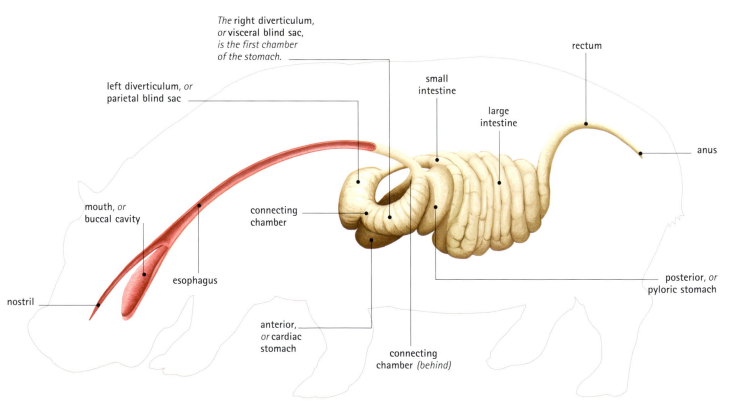

The right diverticulum, *or* visceral blind sac, *is the first chamber of the stomach.*

left diverticulum, *or* parietal blind sac

small intestine

rectum

large intestine

anus

mouth, *or* buccal cavity

connecting chamber

esophagus

nostril

anterior, *or* cardiac stomach

connecting chamber *(behind)*

posterior, *or* pyloric stomach

◄ A pygmy hippo swims underwater in an African river. This species feeds on aquatic plants as well as terrestrial grasses. Both species of hippos have a multi-chambered stomach that is able to process large quantities of plant material.

period of intensive grazing is stored in the stomach until it can be processed. The posterior stomach has a glandular lining that secretes gastric juices, which accelerate the process of digestion.

To the intestines and beyond

Food leaves the posterior stomach via an opening called the pylorus and enters the first part of the intestine, the duodenum. The pylorus is opened and closed like a drawstring bag. It is controlled by a circular sphincter muscle. The intestines are the site of nutrient uptake. Those of hippos are very long, especially the small intestine, which is up to 100 feet (30 m) long. The colon, or large intestine, is only about one-tenth the length of the small intestine. Hippos have no cecum, the structure to which the appendix is attached in all other artiodactyls and many other herbivores. The waste products pass from the colon to the rectum and are excreted through the anus. Hippos use their feces to good effect. They are territorial animals, with a messy but effective

way of staking their claim to a particular patch of land. They mark their territories with deposits of dung. As they defecate, hippos use a vigorous paddling movement of the tail to spatter the sloppy dung over as wide an area as possible. This helps them overpower the smell of neighboring animals.

IN FOCUS

The richness of hippo dung

The feeding habits of common hippos have some interesting ecological implications. Most grazing animals live and feed in much the same area for extended periods, eating plants and then fertilizing the ground with their droppings. In this way they permit the recycling of nutrients. Hippos, however, usually deposit their droppings in rivers or lakes some distance from their feeding areas. The nutrients in their food are not returned to the ground they came from, resulting in an overall nutrient loss. However, one ecosystem's loss is another's gain; the hippo droppings enrich the waters downstream from where they live. Local fishermen often take advantage of the abundant fish stocks that thrive in the nutrient-rich waters below places where hippos wallow.

Reproductive system

CONNECTIONS

COMPARE the age of sexual maturity of a hippo with that of a small mammal such as a **RAT**.

COMPARE the gestation period of a hippo with that of a **RAT**. Hippos have a gestation period of about 240 days, but in brown rats it is only 23 days.

Male common hippopotamuses reach sexual maturity at seven or eight years of age. However, they rarely attain the social dominance needed to defend a territory successfully and mate until they are a little older. Females reach sexual maturity at about nine years and usually mate in their first possible year. From then on, if conditions allow, they will rear one offspring every two years.

▼ MATING
Common hippo
When mating, the female allows the male partner to climb onto her back. The male supports himself with his forelegs.

IN FOCUS

Suckling infants

Young hippos rely exclusively on their mother's milk for their first two months of life, and it continues to form an important part of their diet until they are weaned at eight to ten months. Suckling usually takes place in the water, so the baby hippo has to dive under the water to reach one of the two teats located between its mother's back legs. Interestingly, even when being suckled on land, young hippos automatically close their nostrils and fold back their ears when they attach to the teat. The young have a mobile, muscular tongue, which is vital in latching on to the mother's teat; the lips are too stiff and hard to create a proper seal, so the young hippo sticks out its tongue and folds it around the teat before sucking.

Suckling on land
A young hippo is able to suckle from one of two small nipples. Occasionally, a mother gives birth to twins.

Internal anatomy

In the male, the testes are internal, and the S-shaped penis is usually retracted. It is often difficult to tell males and females apart superficially, although males usually have larger tusks. Behavior is a more reliable means of sexing a hippo than trying to spot clues from its external genitalia. The female reproductive anatomy is fairly typical of a large mammal, with two ovaries connected to the uterus by short fallopian tubes. The uterus is Y-shaped, or bicornuate, with the arms of the Y being longer than the stem. Fertilized eggs usually implant in the arm connected to the ovary from which the egg was released.

The vagina of a female hippo is unusual in having a series of pronounced ridges running around it. The only other mammals in which these ridges are so pronounced are pigs and whales. Biologists do not fully understand the function of the ridges. Perhaps it has something to do with an aquatic lifestyle.

Mating

Mating takes place in the water, with the male mounting the female from behind, and often half-drowning her in his enthusiasm. Gestation lasts about 240 days; births usually happen on land, but occasionally take place in water, especially if the female feels threatened. Baby hippos weigh between 55 and 120 pounds (25 and 55 kg) at birth.

A hippo that survives through infancy has a life expectancy of about 30 years, after which wear and tear on the teeth will begin to take a toll, and feeding becomes more difficult. There are reports of wild hippos reaching the age of 45, but this is exceptional.

AMY-JANE BEER

FURTHER READING AND RESEARCH

Eltringham, S. K. 1999. *The Hippos.* Poyser Natural History: London.
Macdonald, David. 2006. *The Encyclopedia of Mammals.* Facts On File: NY.

▲ *A mother will watch over her offspring for the first two years of its life. The young hippo grows rapidly, suckling on land or, more usually, underwater.*

Honeybee

ORDER: Chiroptera   SUBORDER: Megachiroptera
FAMILY: Apidae   GENUS: *Apis*

Honeybees are social insects that have been domesticated by people for at least 5,000 years. They feed on nectar and pollen, collecting their food from flowers and storing it inside a network of wax compartments, or cells, called a honeycomb. The bees convert the stored nectar into honey. Honeybee colonies contain several thousand female worker bees. Male honeybees, or drones, live for only a short time and play no part in the work of the colony.

Anatomy and taxonomy
Scientists group all organisms into taxonomic groups based largely on anatomical features. Honeybees belong to a group of insects called the hymenopterans. This group also includes wasps, ants, and sawflies.

- **Animals** All animals are multicellular. They rely on other organisms for food. Unlike other multicellular life-forms, animals can move around for at least one phase of their lives.

- **Arthropods** Arthropods have a segmented body that is covered in an external skeleton, or exoskeleton. The arthropods include the insects, spiders, scorpions, centipedes, and crustaceans, such as crabs and lobsters.

- **Hexapods** While some arthropods may have several hundred legs, hexapods have just six. Most hexapods are insects, but noninsect hexapods include springtails.

- **Insects** Insects are six-legged arthropods with a body divided into three main segments: the head, thorax (midbody), and abdomen. The head bears the mouthparts and most of an insect's sense organs. The three pairs of legs stem from the thorax, which also contains many of the vital organs. The abdomen holds most of the digestive and reproductive structures. The Insecta class forms by far the largest class of animals.

- **Endopterygotes** The most successful insect groups are the beetles, butterflies and moths, flies, and hymenopterans (wasps, bees, and ants). Along with a few smaller groups, these insects are referred to as endopterygotes. Adult and larval (young) endopterygotes are very different structurally. There is an inactive stage called the pupa, during which most of the larva's tissues are broken down and the adult structures are built up.

▶ This family tree shows all the major groups of hymenopterans: the bees, wasps, ants, and sawflies. Scientists know of around 115,000 species in the order Hymenoptera, but there are probably many more undiscovered species.

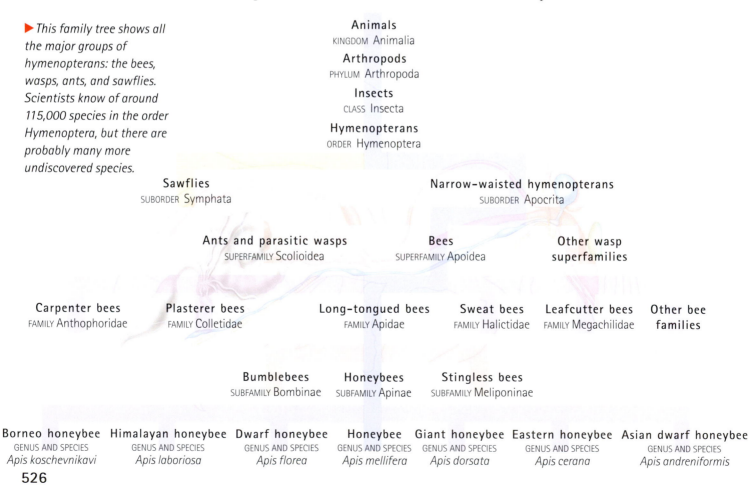

Animals
KINGDOM Animalia
Arthropods
PHYLUM Arthropoda
Insects
CLASS Insecta
Hymenopterans
ORDER Hymenoptera

Sawflies
SUBORDER Symphata

Narrow-waisted hymenopterans
SUBORDER Apocrita

Ants and parasitic wasps
SUPERFAMILY Scolioidea

Bees
SUPERFAMILY Apoidea

Other wasp superfamilies

Carpenter bees
FAMILY Anthophoridae

Plasterer bees
FAMILY Colletidae

Long-tongued bees
FAMILY Apidae

Sweat bees
FAMILY Halictidae

Leafcutter bees
FAMILY Megachilidae

Other bee families

Bumblebees
SUBFAMILY Bombinae

Honeybees
SUBFAMILY Apinae

Stingless bees
SUBFAMILY Meliponinae

Borneo honeybee
GENUS AND SPECIES
Apis koschevnikavi

Himalayan honeybee
GENUS AND SPECIES
Apis laboriosa

Dwarf honeybee
GENUS AND SPECIES
Apis florea

Honeybee
GENUS AND SPECIES
Apis mellifera

Giant honeybee
GENUS AND SPECIES
Apis dorsata

Eastern honeybee
GENUS AND SPECIES
Apis cerana

Asian dwarf honeybee
GENUS AND SPECIES
Apis andreniformis

● **Hymenopterans** These insects have one pair of large forewings and another smaller pair behind. These wings may be lost at some point in the life cycle. In flight, the hind wing clips onto the forewing to make a single flying surface. The Hymenoptera includes sawflies, wasps, ants, and bees. Some hymenopterans, including honeybees, yellow jackets, and ants, live in colonies, but most species live solitary lives.

● **Apocritans** The suborder Apocrita includes all ants wasps, and bees. Members of this group have a narrow "waist" between the first and second subsegments of their abdomen. This makes the abdomen very flexible.

● **Bees** The superfamily Apoidea groups all the bees together. Bees feed only on pollen and nectar and do not hunt prey as other apocritans do. Like some other members of the Apocrita, bees have a stinger in place of an ovipositor (egg-laying tube).

● **Long-tongued bees** The Apidae is a family of bees with a long, flexible tongue and pollen baskets on their hind legs. These bees live in large colonies for at least some part of their lives. Many of the colonies are complex, with a single queen bee producing all the young. Her daughters then work together to rear more young and maintain the nest.

● **Bumblebees** Bumblebees are large and robust. Those that live in North America and other mild climates are generally covered in dark hairs. Their tropical relatives are less hairy and come in a range of shiny colors. Bumblebees are generally less social than other members of the Apidae.

● **Stingless bees** Stingless bees are small and live mainly in tropical regions. They are unable to use their small stinger and instead bite attackers. Stingless bees live in large

▲ Worker honeybees are responsible for collecting nectar and depositing it in honeycombs they construct from wax.

colonies. Instead of nectar, some stingless bees collect plant resins; others gather honeydew, a sugary liquid produced by sap-sucking bugs.

● **Honeybees** Honeybees are one of seven species in the genus *Apis*. Although all members of this genus produce beeswax and honey, only the honeybee can produce it in quantities large enough to be worth collecting by people. Honeybees and other *Apis* bees live in colonies of up to 80,000 individual bees.

FEATURED SYSTEMS

EXTERNAL ANATOMY Honeybees are narrow-waisted insects with two pairs of thin wings. They have hairs on their hind legs that collect pollen grains, and a stinger at the tip of the abdomen. *See pages 528–531.*

INTERNAL ANATOMY Honeybees have many glands inside their bodies. These glands produce a range of substances from venom to chemical signals called pheromones. *See pages 532–533.*

NERVOUS SYSTEM Honeybee eyes are sensitive to ultraviolet light, which is invisible to humans and other mammals. In addition to two compound eyes, honeybees have three eyes on top of their head. These insects are also very sensitive to touch and chemical signals. *See page 534.*

CIRCULATORY AND RESPIRATORY SYSTEMS Like other insects, honeybees breathe through holes in the exoskeleton called spiracles. Oxygen travels into the body through the spiracles and along tubes called tracheae. The heart is part of a tubelike structure called the dorsal vessel. *See page 535.*

DIGESTIVE AND EXCRETORY SYSTEMS Honeybees have a sac called a honey stomach in their abdomen, which they use to store nectar before turning it into honey back at the hive. *See page 536.*

REPRODUCTIVE SYSTEM Most honeybees do not reproduce. Instead they work together to help a single queen bee, their mother, produce thousands of offspring. *See page 537.*

External anatomy

COMPARE the wings of the honeybee to those of a *DRAGONFLY*. All four of the bee's wings move up and down at the same time. In dragonflies, the hind pair move up as the forewings move down.

COMPARE the honeybee's mouthparts with those of a *WEEVIL*. These beetles have long tubelike snouts, which they use to eat plants.

Like all insects, a honeybee has a body arranged in three sections: the head, the thorax, and the abdomen. The head carries the bee's mouthparts and main sensory organs, such as eyes and antennae. The head is connected to the thorax, or midbody, to which six legs and four wings are attached. The narrow waist that is characteristic of bees, wasps, and ants is often thought to be located between the thorax and the final body section, the abdomen. The waist actually forms between the first and second subsegments of the abdomen. The first abdominal section is fused with the thorax behind the point at which the wings are attached.

Three forms

Honeybees come in three forms, or castes: workers, drones, and queens. All the castes follow the same basic body plan but have a few different anatomical features. The honeybees seen on flowers and in flight are worker bees. Workers are smaller than drones or queens. All workers are female and have a stinger on their abdomen. The stinger can be used only once. It detaches into the skin of an enemy, tearing out some of the bee's internal organs with it so that the insect dies soon after.

Drones are a little larger than workers. Honeybee drones are male. They never have a stinger, and, unlike workers, drones can mate.

▼ *The honeybee's body consists of a head, thorax, and abdomen. Wings and legs are attached to the thorax, while the larger abdomen contains important internal organs.*

0.6 inch (2.2 cm)

The **abdomen** *contains most of the honeybee's internal organs and, despite appearances, begins just forward of the bee's narrow waist. The abodomen is divided into seven subsegments.*

wings

The **thorax** *contains the strong flight muscles that power the bee's wings. On honeybees, both the thorax and the* **abdomen** *are covered in heat-conserving hairs.*

basitarsus

middle tarsus segments

tarsus

pretarsus

The **compound eye** *is made of thousands of tiny light-sensitive cells called* **omatidia.**

The honeybees's **antennae** *are jointed in the middle and bend downward. Fine hairs on the antennae detect chemicals in the air.*

Wing anatomy

Honeybees, like all hymenopterans, have two pairs of thin wings. The wings are made from a delicate membrane (the name hymen-optera means "membrane wing"). The fore-wings are larger than the hind wings. A row of tiny hooks called hamuli lie on the front edge of the hind wings. They attach to a fold on the back edges of the forewings. This connection makes the two wings function as a single wing. Honeybees can reach speeds of 15 miles per hour (24 km/h), but they can fly for only about 15 minutes before they run out of energy and need to eat something.

However, they die soon afterward because they cannot feed themselves. Their tongue is much shorter than that of the female, so they cannot reach into flowers to suck up nectar. The largest honeybees are the queens. Each colony has only one queen, who is the mother of all the workers. A queen is almost twice the size of a worker.

A hard body

All insects, including honeybees, have an exoskeleton. The exoskeleton covers the whole body, protecting the soft organs and tissues inside. It also gives the body a rigid structure in the same way as an internal bony

Each **compound eye** is very large.

The **thorax** is broad.

Drone

The *abdomen is broad and segmented.*

The queen has a longer **abdomen** *than workers and drones.*

Queen

▲ Queen and drone honeybee

The queen bee is the largest of the three types of honeybees, and there is only one queen in a hive. She lives for around two years and may lay 1,500 eggs per day. The drone's only function is to mate with a queen, after which he dies.

Hairy legs

One of the jobs of a worker honeybee is to collect pollen from flowers. Pollen is an important source of fats and proteins used by developing bee larvae (young). Many pollen grains attach to the bee's hairy body when it lands on the flower, but honeybees also have dedicated pollen-collection structures called pollen baskets on their legs. These are loaded with pollen before the bee

flies back to the nest. Each basket is a concave area on the outside of the tibia of the hind leg, surrounded by long, curved hairs. The honeybee uses stiff bristles on the underside of its other legs to comb pollen grains out of its body hair with a grooming motion. It then transfers this collected pollen from leg to leg and into the baskets. Once back at the nest, the honeybee kicks the

pollen off its legs into an empty cell. A cell is a single six-sided unit of a honeycomb. The honeycomb is made from beeswax and is used to hold pollen, honey, and developing bee larvae.

Other types of bees collect pollen in hairy patches located on the legs or abdomen, but unlike honeybees and other long-tongued bees they do not have pollen baskets.

skeleton. The exoskeleton is made from a tough but flexible material called chitin. All the body's moving parts, such as legs and other appendages, are arranged in jointed segments so they can bend and rotate easily despite being encased in a rigid covering.

Striped for danger

Honeybees are covered by hairs. These have several functions. They keep the bee warm, trapping the heat produced by the large flight muscles. This allows honeybees to fly when the temperature is too cold for most other insects. The hairs are also sensitive to touch. This sense is very important for communication in the darkness of the nest interior. Worker bees also use their hairs to collect pollen, an important food source for young honeybees.

The hairs on the abdomen of a worker honeybee form yellow and black stripes. This pattern also occurs on some other bees and on social wasps. The stripes act as a warning to predators that the insects can sting. Although this warning might not save an individual bee, the predator will remember that eating a yellow and black insect is painful and best avoided in future. Some insects without stingers, such as bee-mimic moths and hoverflies, have yellow and black stripes on their abdomen. The mimics rely on this deception to keep predators at bay.

Africanized bees

During the last 50 years, North and South America have seen the arrival of so-called Africanized, or killer, bees. These bees appeared

---

CLOSE-UP

Tongue twisters

The long tongue, or glossa, of an adult honeybee is used for sucking up liquids, such as nectar, honey, or water. The tongue is a very flexible organ. It can be lengthened or shortened and moved in any direction. When fully extended, the tongue is about the same length as the bee's head. The long tongue helps honeybees suck up nectar from deep within flowers.

The tongue is flanked by two pairs of appendages called palps. Since the adults do not eat solid foods, these mouthparts are used to handle pollen and wax. Higher up on the head, a pair of pincerlike mandibles is also used for handling wax during construction of honeycombs.

Tongue and mouthparts
Unlike the mouthparts of many other insects, such as those of grasshoppers and moths, the mouthparts of the honeybee allow it to both chew and suck.

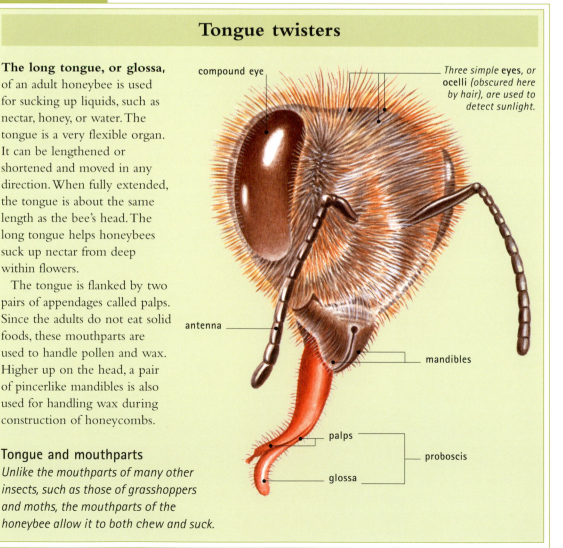

compound eye

Three simple eyes, or ocelli (obscured here by hair), are used to detect sunlight.

antenna

mandibles

palps

proboscis

glossa

COMPARATIVE ANATOMY

A passenger compartment

Mites are a problem for many types of bees, including honeybees. These tiny relatives of spiders live as parasites in bees' nests, sometimes spreading diseases that kill entire colonies. However, not all mites are bees' enemies. Some giant carpenter bees rely on mites to keep their larvae clean and free of fungal infections. Carpenter bees are so called because they dig their nest holes into wood. When a female carpenter bee becomes adult and is ready to leave the nest, she places some mites on her body. The mites scurry into a pouch in the bee's abdomen just behind her narrow waist. When the bee finishes making a new nest, she releases the mites. They then feed on the skin shed by developing bee young and thus keep the nest clean.

Tiny **mites** *are carried within a pouch in a female carpenter bee's abdomen. The mites will help to keep her nest clean.*

◀ *The hairs on a honeybee's body become covered with pollen as it sucks nectar from a flower. Some of this pollen is carried by the honeybee back to the hive, but some falls off on the other flowers that the bee visits. The pollen that is left behind may then fertilize a flower's ova.*

when researchers released African honeybees into the wild in Brazil by mistake. The African bees mated with the local honeybees, producing Africanized bees. The anatomy of Africanized bees is almost identical to that of other honeybees. The main difference is their aggressiveness. Africanized bees will sting attackers in large swarms much more readily than other types of honeybees. Although it is possible for people to die if they suffer several hundred bee stings, Africanized bees very rarely kill.

Internal anatomy

CONNECTIONS

COMPARE the way a honeybee stinger works with that of a *JELLYFISH*. These animals have tiny stinging cells that contain coiled threads armed with venom-bearing barbs.

As with all arthropods, a honeybee's muscles are inside the exoskeleton. The muscles are attached to struts that stick out from the inside of the exoskeleton. The muscles pull against these structures, in the same way that vertebrate muscles pull against bones. Honeybees have especially strong muscles in the throat that help them suck up fluids.

Glands in the head

Honeybees have many glands inside their body. A gland is an organ that produces a particular substance. Salivary glands in the head of most insects produce saliva, a liquid that softens food and starts the digestive process. Adult honeybees do not use saliva in this way because they feed on liquid food. However, they have two salivary glands, one in the head and the other in the thorax. There are two other main glands in the head, the hypopharyngeal gland and the mandibular gland. Although biologists are not sure exactly how, these four glands together produce a substance called royal jelly, a milky liquid that workers feed to larvae.

Abdominal glands

The glands in the abdomen of most female insects produce liquids that coat eggs as they are laid. This coating has many functions. It may keep the eggs moist, help them stick in the right place, or make it hard for killer parasites to get inside the egg. In honeybee workers and other female bees, this gland is called Dufour's gland for the French biologist who discovered it.

Most female bees use their Dufour's gland to line their nests. The waxy secretions make the nest watertight. However, Dufour's gland has a different function in honeybees. The queen has

▶ Since the honeybee's body narrows to a very slender waist, the nervous system, digestive system, and circulatory system must all bunch together at this point. The large abdomen provides space for many of the honeybee's internal organs and protects them inside the hard exoskeleton.

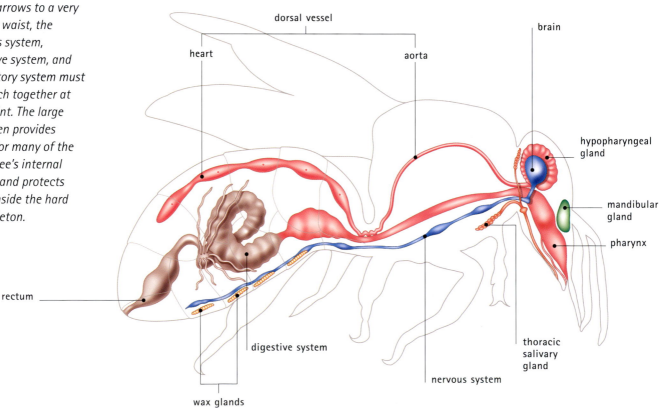

dorsal vessel

brain

heart

aorta

hypopharyngeal gland

mandibular gland

pharynx

rectum

digestive system

thoracic salivary gland

nervous system

wax glands

CLOSE-UP

A stinger in the abdomen

The honeybee stinger evolved from the ovipositor, the tube at the tip of the abdomen through which other female insects lay eggs. Only female bees sting. Males do not lay eggs and so have no ovipositor. Worker honeybees do not lay eggs, but queen bees do. Since their ovipositors are used as stingers, queens release eggs through a hole at the base of the stinger instead.

A worker honeybee's stinger is used only once. When a worker honeybee jabs its stinger into an attacker, barbs on the tip become lodged in the victim's body. The stinger and venom gland detach from the bee's body. Muscles in the stinger keep working to pump in the venom. The worker bee dies soon after. Queen honeybees have a smooth stinger that can be used repeatedly. They use it for killing rival queens.

Stinging organ
The stinger consists of two barbed lancets. After the stinger has detached from the bee, muscles in the stinger continue to move these lancets alternatively, working the sting deeper into the victim.

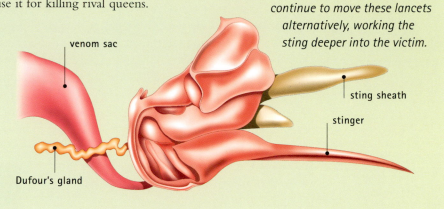

venom sac

Dufour's gland

sting sheath

stinger

---

a larger gland than her workers. She uses it to coat each egg with a waxy liquid. This liquid is a signal to workers not to dispose of the eggs during their regular cleanouts.

Glands for wax

Worker honeybees build honeycombs out of beeswax. They produce beeswax with wax glands located along the underside of their abdomen. A worker's wax glands develop a couple of days after the insect emerges as an adult. The glands convert sugars from honey into wax. The wax oozes out through pores on the abdomen, where it forms white flakes. The bee chews up the flakes and adds them to the honeycomb it is building. After a couple of weeks, the worker is too old to make wax efficiently and starts doing other work.

IN FOCUS

Flight muscles

The flight muscles of honeybees and most other flying insects are not attached directly to the wings. Instead the muscles move the wings by changing the shape of the thorax. Two sets of muscles are involved. One set runs from the top of the thorax to the bottom. When these muscles tighten, the thorax is compressed and the wings rise up. The other set runs the length of the thorax. When they contract, the thorax changes shape, and the wings are pulled down again.

◄ *The honeybee's powerful flight muscles enable it to hover over flowers and position itself for landing.*

Nervous system

Honeybees have all the same senses that people have, and a few more besides. They have five eyes, three simple ones (ocelli) on top of the head and two large compound eyes on each side of the head. The ocelli are little more than light-sensitive patches that tell the bee the position of the sun. Compound eyes are made up of several thousand individual units called ommatidia. Each ommatidium has its own lens and light-sensitive nerve cell. Compound eyes are good for detecting movement, but they do not form sharp, focused images. Bees are also sensitive to ultraviolet light, which is invisible to mammals. However, they cannot see the color red at all.

Hairs and senses

As with most insects, hairs are important features of a honeybee's nervous system. Hairs around the mouthparts are used to taste foods, and vibration-sensitive hairs all over the body can detect sounds. Hairs such as those on the antennae are especially important for touch. Like the antennae of most hymenopterans, a honeybee's two antennae are each jointed in the middle to form an "elbow." Honeybees have a notch on each foreleg called an antenna cleaner. Running the antennae through the notch keeps them in good condition.

Hairs on the antennae are able to detect chemicals in the air as part of the bee's sense of smell. The antennae are also used as feelers. A honeybee's sense of touch is especially useful when a worker is dancing. A bee dances to tell other workers where they can find food. The other bees gather around the dancer and feel her movements, which tell them in which direction and how far to travel.

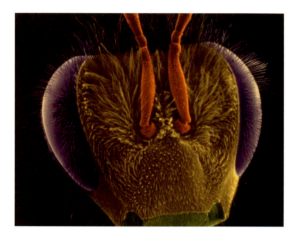

▲ A bee's compound eyes are made up of thousands of individual light-sensitive cells called ommatidia.

▶ The nervous system of a honeybee is similar to that of other insects. The nerve cord runs ventrally (close to the underside) of the body and loops around the esophagus to the brain. Along the length of the nerve cord, nerves branch from bunches of nerve cells called ganglia.

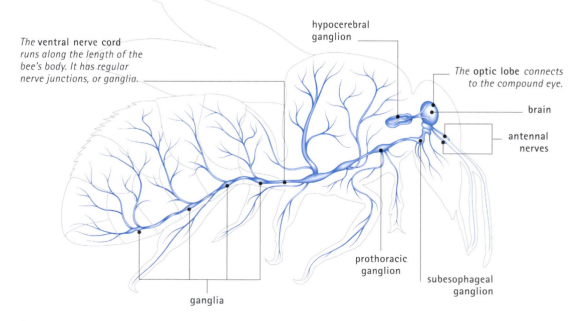

The **ventral nerve cord** runs along the length of the bee's body. It has regular nerve junctions, or ganglia.

hypocerebral ganglion

The **optic lobe** connects to the compound eye.

brain

antennal nerves

prothoracic ganglion

subesophageal ganglion

ganglia

534

Circulatory and respiratory systems

The circulatory and respiratory systems of honeybees are typical of other insects. The circulatory system carries digested food to all parts of the bee's body. The respiratory system delivers oxygen to the body's tissues and removes carbon dioxide.

Hemolymph and tracheae

Insects, unlike vertebrates, do not have blood running through veins and arteries. Instead, everything inside an insect's body is bathed in a yellow bloodlike liquid called hemolymph. The hemolymph is circulated using a large tube called the dorsal vessel that runs along the length of the body. The rear section of this tube runs under the top side of the abdomen. This section acts as a pump and is called the heart. The heart moves the hemolymph toward the front of the body. The forward section of the dorsal vessel does not pump and is called the aorta. The aorta carries hemolymph to the brain. Hemolymph passes into the body cavity, or celom. It passes back into the heart through small holes called ostia that run along its side. Valves prevent fluid from leaving the heart through these holes.

IN FOCUS

Temperature control

Honeybee larvae grow best at a temperature of around 95°F (35°C). In cold conditions, worker honeybees warm up the nest by crowding together and shivering with their flight muscles. The heat produced by the thousands of muscles warms the nest. If it is too hot, workers stand at the entrance to the hive and fan their wings to draw in cool air. Others may suck up water and spray it around the nest to cool their nest mates.

Arthropods do not have lungs, and insects do not breathe through their mouth. Instead, adult insects get their oxygen through small holes in the exoskeleton called spiracles. The spiracles are the openings of a system of tubes called tracheae, which connects the inside of the insect's body with the air outside. Oxygen in the air passes into the tracheae and deeper into the body through smaller tubes called trachioles. Eventually the oxygen diffuses directly into the tissues that need it. Carbon dioxide, a waste product of metabolism, travels in the opposite direction out into the air.

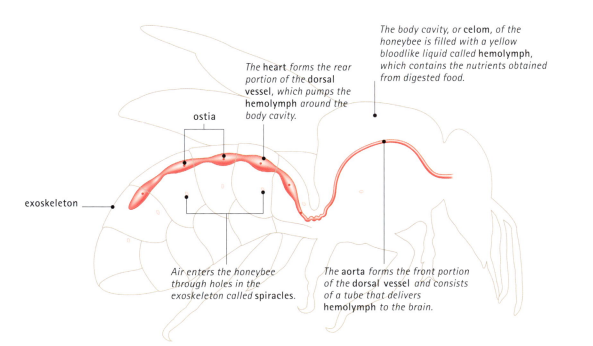

The **heart** forms the rear portion of the **dorsal vessel**, which pumps the **hemolymph** around the body cavity.

ostia

The body cavity, or **celom**, of the honeybee is filled with a yellow bloodlike liquid called **hemolymph**, which contains the nutrients obtained from digested food.

exoskeleton

Air enters the honeybee through holes in the exoskeleton called **spiracles**.

The **aorta** forms the front portion of the **dorsal vessel** and consists of a tube that delivers **hemolymph** to the brain.

◀ CIRCULATORY SYSTEM

In contrast to vertebrates such as humans, insects such as honeybees have an open circulatory system. The insect equivalent of blood, called hemolymph, is pumped by a simple heart directly into the body cavity. In vertebrates the blood is pumped through a closed system of arteries and veins.

Digestive and excretory systems

Nectar is a sugary liquid produced by flowers. It is an excellent source of energy. Honeybees use nectar to make honey, a gooey sweet liquid that adult honeybees feed on. Honeybee larvae eat a combination of honey and pollen. Honey and pollen are mixed together by workers called nurse bees, making a food called "bee bread."

Food processing

Honeybees carry pollen back to the nest in hairy pollen baskets on their legs. The pollen is stored until it is fed to a larva. Foraging workers suck up nectar from flowers into a sac in their abdomen called the honey stomach. In the honey stomach, the complex sugars in the nectar are broken down into simpler substances such as glucose. Back at the nest, the bee regurgitates the digested nectar from its honey stomach out of its mouth and into a cell. The water in the nectar begins to evaporate, causing the liquid to thicken and become honey.

Since they feed only on liquid food, adult honeybees need powerful sucking muscles in their throats. The same muscles are also used for regurgitating fluids. During feeding, food

passes beyond the honey stomach and is digested in the midgut and hindgut.

Insects remove wastes from their body fluids with organs called Malpighian tubules. The tubules float in the celom and are surrounded by hemolymph. They attach to the digestive system at the junction of the midgut and hindgut. Wastes are released into the hindgut and leave the body through the anus.

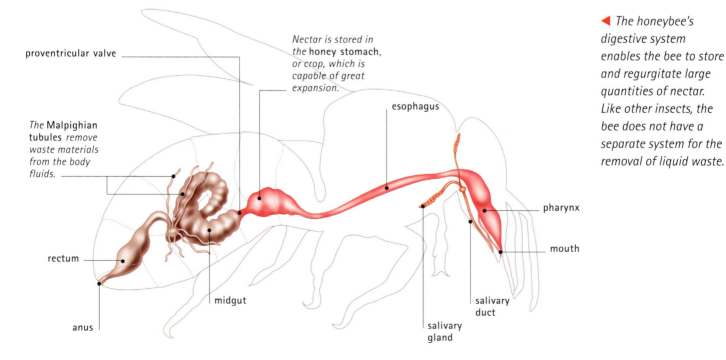

proventricular valve

Nectar is stored in the **honey stomach,** or crop, which is capable of great expansion.

esophagus

The **Malpighian tubules** *remove waste materials from the body fluids.*

rectum

anus

midgut

salivary gland

salivary duct

mouth

pharynx

◀ *The honeybee's digestive system enables the bee to store and regurgitate large quantities of nectar. Like other insects, the bee does not have a separate system for the removal of liquid waste.*

Reproductive system

Like many insects, honeybees undergo complete metamorphosis. That is, they change from one body form to a completely different one as they become adults. Whether an egg grows into a worker bee, a drone, or a queen depends on how it is raised by the workers that nurse it. Drones grow from eggs that do not get fertilized with sperm by the queen. With half the normal number of genes, drones are said to be haploid. Worker bees and queens are diploid; they develop from eggs that have been fertilized.

Larval life

Honeybee eggs are the size of a pinhead. A tiny wormlike larva hatches out about three days after the egg is laid. For the first few days of its life, the larva eats a creamy liquid (called royal jelly), which is produced by its nurses. Most larvae are then fed a mixture of honey and pollen. However, if a larva is fed nothing but royal jelly, it will become a queen bee. Queens are nursed in extra-large brood cells.

The larva develops through a series of stages called instars. It sheds its exoskeleton from time to time to allow growth to occur. Each molt marks the change from one instar to another.

Becoming adult

After its fifth molt, the larva becomes a pupa. The nurse bees seal the pupa into its cell with wax. Metamorphosis now takes place. Worker

▶ FEMALE REPRODUCTIVE ORGANS

The female's sex organs are closely connected to the sting apparatus. The sting evolved from a structure called an ovipositor, which was used to lay eggs.

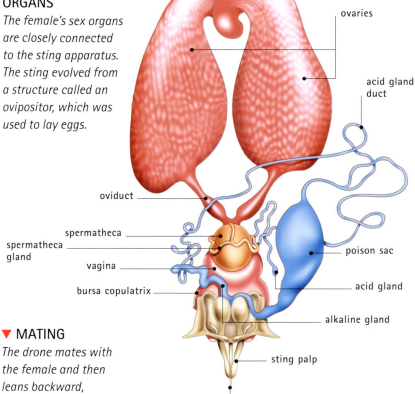

ovaries

acid gland duct

oviduct

spermatheca

spermatheca gland

vagina

bursa copulatrix

poison sac

acid gland

alkaline gland

sting palp

terebra

▼ MATING

The drone mates with the female and then leans backward, causing his endophalus to snap off in the queen. Soon after this the drone dies. The queen stores some of the drone's sperm in the spermatheca and goes on to mate with several other males.

pupae take 21 days to turn into adults; drones take 24 days to develop. Adult drones fly away from the nest looking for queens with which to mate. When the resident queen dies, the workers make new ones to take her place.

As soon as they emerge, the queens sting to death any emerging rivals until only one queen remains. She leaves the nest to mate with several drones from other colonies. The queen comes back laden with enough sperm to lay about 2,000 fertilized eggs every day for up to five years.

TOM JACKSON

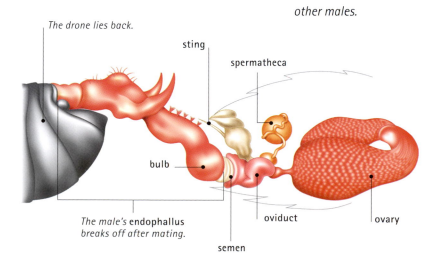

The drone lies back.

sting

spermatheca

bulb

The male's **endophallus** breaks off after mating.

semen

oviduct

ovary

FURTHER READING AND RESEARCH

Austin, Andrew, and Mark Dowton. 2000. *Hymenoptera: Evolution, Biodiversity and Biological Control.* CSIRO Publishing: Melbourne, Australia.

Green, Rick. 2002. *Apis Mellifera: A.K.A. Honeybee.* Branden Publishing Co.: MA.

Housefly

SUBCLASS Insecta ORDER: Diptera
FAMILY: Muscidae SPECIES: *Musca domestica*

Houseflies are small, flying insects that are often seen inside people's houses. They occur worldwide and often live near rotting vegetation and garbage, on which female houseflies lay their eggs.

Anatomy and taxonomy

Biologists classify organisms into taxonomic groups based largely on features of their anatomy. In insects, characteristics of the mouthparts, wing veins, and bristles on the body are often important for identifying individual species. The housefly belongs to the family Muscidae. Along with dung flies, blowflies, hoverflies, and a number of similar groups, they form part of a larger group of flies called the Cyclorrhapha.

● **Animals** All animals are multicellular and depend on other organisms for food. They differ from other multicellular life-forms in their ability to move around (generally using muscles) and to respond rapidly to stimuli.

● **Arthropods** Arthropods are animals with segmented bodies, although segments are often fused to form units such as the head and abdomen. They have a tough outer skin called an exoskeleton that protects the internal organs and serves as an attachment point for muscles. To grow, an arthropod must molt its exoskeleton.

Arthropods have pairs of jointed appendages, modified to form structures such as legs, mouthparts, and antennae. Internally, all arthropods have a ventral (running along the underside) nerve cord and a main dorsal (running near the top) blood vessel. The dorsal vessel pumps a liquid called hemolymph around the body cavity (or hemocoel).

● **Insects** Insects are hexapods, six-legged arthropods. All hexapods have three pairs of segmented legs, and their bodies are organized into a head, thorax, and abdomen. Noninsect hexapods do not have wings or antennae, and are further distinguished from the insects by the structure of their mouthparts, which are kept in a pouch. The insects display the most diverse body forms and functions of all the classes of animals on Earth.

● **Endopterygotes** The largest and most successful insect groups are the beetles, butterflies and moths, flies, and hymenopterans (wasps, bees, and ants). Along with a few smaller groups, these insects are referred to as

▼ *This tree shows the major groups to which houseflies belong. Note that many insects have the word* fly *in their name, such as dragonflies, caddis flies, and scorpion flies. However, true flies belong to just one order, the Diptera.*

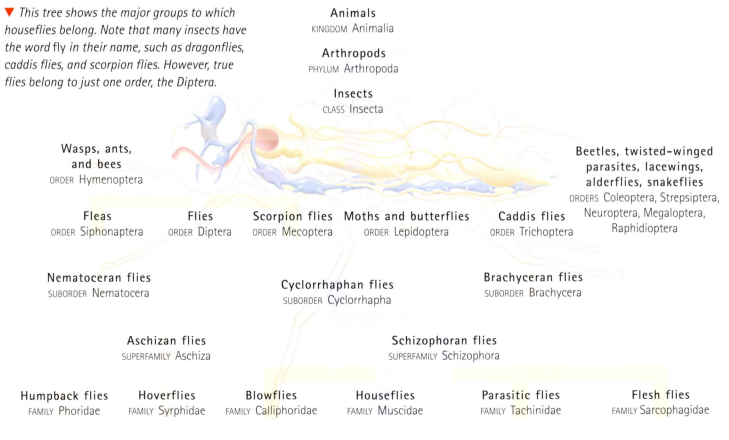

Animals
KINGDOM Animalia

Arthropods
PHYLUM Arthropoda

Insects
CLASS Insecta

Wasps, ants, and bees
ORDER Hymenoptera

Beetles, twisted-winged parasites, lacewings, alderflies, snakeflies
ORDERS Coleoptera, Strepsiptera, Neuroptera, Megaloptera, Raphidioptera

Fleas
ORDER Siphonaptera

Flies
ORDER Diptera

Scorpion flies
ORDER Mecoptera

Moths and butterflies
ORDER Lepidoptera

Caddis flies
ORDER Trichoptera

Nematoceran flies
SUBORDER Nematocera

Cyclorrhaphan flies
SUBORDER Cyclorrhapha

Brachyceran flies
SUBORDER Brachycera

Aschizan flies
SUPERFAMILY Aschiza

Schizophoran flies
SUPERFAMILY Schizophora

Humpback flies
FAMILY Phoridae

Hoverflies
FAMILY Syrphidae

Blowflies
FAMILY Calliphoridae

Houseflies
FAMILY Muscidae

Parasitic flies
FAMILY Tachinidae

Flesh flies
FAMILY Sarcophagidae

endopterygotes. Adult and larval (young) endopterygotes are very different structurally. There is an inactive stage called the pupa, during which most of a larva's tissues are broken down and the adult structures are built up.

● **Flies**   True flies form the insect order Diptera. Like beetles, they use just one pair of wings for flight. In place of the second pair of wings, dipterans have two club-shaped structures called halteres. They help the insect balance in flight. All dipterans feed primarily on liquid food, but there is great variation in the form of their mouthparts—from the piercing mouthparts of mosquitoes to the fleshy, spongelike mouthparts of blowflies. Biologists divide the Diptera into three suborders—the Cyclorrhapha, Nematocera, and Brachycera.

● **Nematoceran flies**   Members of this suborder have slender bodies and long antennae. They include the crane flies, mosquitoes, midges, and blackflies.

● **Brachyceran flies**   These flies have shorter antennae and are stouter than the Nematocera. The brachycerans include the horseflies and robber flies. Brachycerans are often confused with cyclorrhaphan flies, but adults leave a straight or T-shaped split in their pupal case when they emerge as adults.

● **Cyclorrhaphan flies**   The suborder Cyclorrhapha contains the most advanced families of flies. Members of this group are distinguished from those of the other two suborders, Nematocera and Brachycera, by the circular hole that the adult makes when it emerges from its pupal case. In addition to houseflies, the Cyclorrhapha includes the hover flies, many of which look like two-winged bees and wasps; the tiny fruit flies with their brightly colored eyes; flesh flies and blowflies, the larvae of which feed on dead or rotting flesh; and the warble flies and botflies, whose larvae feast on living flesh.

▲ *This housefly is preparing to feed on some bread. The large, red compound eyes take up much of the head of the fly.*

● **Schizophoran flies**   The Schizophora are distinguished from the aschizan flies by the presence of a prominent groove, the ptilinal suture, on the head.

● **Houseflies**   Flies from the family Muscidae are small, and are distinguished from other schizophoran flies by the absence of a fan of bristles on their thorax and by the presence of feathery aristae. An arista is the end section of an antenna. Houseflies have spongelike mouthparts and must liquefy their food before sucking it up. Other muscids, such as the stable flies, feed by biting mammals and drinking their blood.

FEATURED SYSTEMS

EXTERNAL ANATOMY   Houseflies are small flying insects with one pair of wings and one pair of balance organs called halteres. Their body is covered in a tough, waterproof exoskeleton. *See pages 540–543.*

INTERNAL ANATOMY   Flies need powerful muscles to operate their wings during flight. The main body cavity, the hemocoel, contains a bloodlike fluid called hemolymph that bathes the internal organs and supplies them with nutrients. *See page 544.*

NERVOUS SYSTEM   A fly's ventral nerve cord consists of ganglia that receive impulses from the sense organs such as the compound eyes and taste receptors. *See page 545–546.*

CIRCULATORY AND RESPIRATORY SYSTEMS   Airways called tracheae permit gaseous exchange in the thorax and abdomen of adult flies. Fly larvae have various structures to help them breathe in low-oxygen environments. *See page 547.*

DIGESTIVE AND EXCRETORY SYSTEMS   Flies digest their food externally before sucking it into the gut through their spongelike mouthparts. Excretion of nitrogenous waste occurs through the Malpighian tubules. *See page 548.*

REPRODUCTIVE SYSTEM   Flies reproduce sexually by internal fertilization. Most females lay eggs that hatch into larvae, which undergo a radical transformation during a life stage called pupation to become adults. *See page 549.*

External anatomy

An adult common housefly is about 0.3 inch (0.7 cm) long and has a wingspan of up to 0.6 inch (1.5 cm). Its hairy body is gray-yellow in color, with four dark stripes along its back. As with all other adult insects, a housefly's body is made up of three main sections. They are the head, the thorax (or midbody), and the abdomen.

The housefly's head is formed from six body segments that are fused into a hard capsule. The main external structures of the head are a pair of large compound eyes; simple eyes called ocelli; and the sensitive feelers, or antennae. Most cyclorrhaphan flies have a join on the top of their head called the ptilinal suture. This forms during the emergence of the adult fly from its puparium. This is the tough case inside which the larva (young) changes into an adult, a transition called complete metamorphosis. A balloonlike sac called the ptilinum inflates to force open the puparium, allowing the adult fly to escape. After emergence from the puparium, the ptilinum deflates. The ptilinal suture then closes over it.

The head and thorax of flies are covered by many sets of bristles. Entomologists (scientists who study insects) examine the arrangement of the bristles to identify different fly species.

Moplike mouthparts

Like all insects, flies have external mouthparts. The main features that occur in all flies are the labrum (equivalent to an upper lip), labium (equivalent to a lower lip), and hypopharynx, which contains the salivary duct along which saliva passes. These structures take

▼ *The modified hind wings, or halteres, on the metathorax distinguish flies from all other insects.*

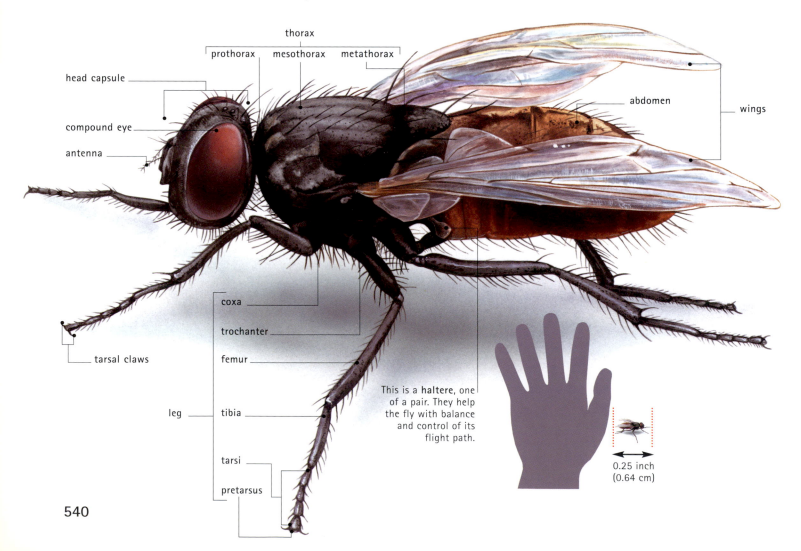

thorax
prothorax mesothorax metathorax
head capsule
compound eye
antenna
abdomen
wings
coxa
trochanter
tarsal claws
femur
leg tibia
tarsi
pretarsus

This is a **haltere**, one of a pair. They help the fly with balance and control of its flight path.

0.25 inch (0.64 cm)

different shapes depending on their function in different species. A housefly uses its mouthparts to suck up liquid. The labrum and labium together form a fleshy structure called a proboscis. This has a flattened base called a labellum. Tiny tubes called pseudotrachae permeate the labellum. The fly sucks up its liquid food through these tubes.

The thorax

As in all insects, the thorax, or midbody, is subdivided into three sections. They are the prothorax, mesothorax, and metathorax. Each thorax section bears one pair of jointed legs. The main parts of the leg are the coxa and trochanter nearest the thorax, the femur and tibia in the middle, and a series of tarsi at the end. The leg ends in a claw-tipped structure called the pretarsus. The femur and tibia are covered in rows of stiff bristles that the fly uses to clean its eyes, its mouthparts, and other parts of its body.

All dipterans have one pair of membranous wings mounted on the large mesothorax. Scientists often use the patterns of the wing veins to tell one species from another. On the metathorax are the halteres. These structures occur only in some kinds of insects. They

occiput

compound eye

clypeus

The base of the antenna, or scape, contains sensory cells called scolophores. These detect movements of the antennae, and allow the insect to hear sound vibrations.

The maxillary palps contain cells that detect chemicals in the air. This helps the mosquito find a meal. The cells may detect carbon dioxide breathed out by a host. They may also detect chemicals in sweat, or other body odors.

The labrum (upper "lip") and labium (lower "lip") form a sheath that protects the stylets.

The labella act as a guide to the piercing sections of the stylets.

The stylets end in a sharp point. This is used to pierce the skin of the host.

The paired mandibles and maxilla are long, forming structures called the stylets. They consist of two tubes. Saliva passes down one tube, the hypopharynx. Blood passes up the other, the epipharynx.

Hairs on each antenna detect sound vibrations in the air, passing signals to the base of the antenna. This is a female mosquito. Male mosquitoes have much bushier antennae. They listen for the high-frequency whirring of females in flight. They then try to intercept and mate with them.

The maxillae are lined by backward-pointing teeth. They cause damage inside the host, leading to more blood loss than a smooth tip alone would manage.

COMPARATIVE ANATOMY

Sucking insects

Insects eat many different types of food. Adult dipterans feed mostly on liquids. The housefly sponges up fluid from the surface of its food through its fleshy proboscis. Mosquitoes feed on animal blood; they have needlelike mouthparts that can pierce the skin and suck up blood like a straw. Butterflies and moths, which feed on liquid nectar (a sweet liquid) from flowers, also need to suck their food. They have a coiled proboscis that can be extended and retracted by changing the pressure of fluid in the hemocoel (body cavity).

▶ MOUTHPARTS, SIDE PROFILE
Housefly
Flies dribble saliva onto their food before feeding. This breaks it down into a liquid, which they suck through tubes in the labellum.

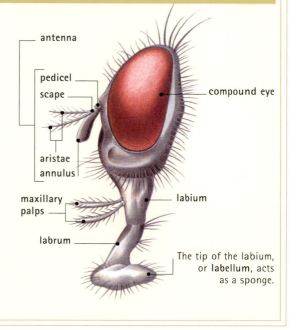

antenna
pedicel
scape
aristae
annulus
maxillary palps
labrum
compound eye
labium
The tip of the labium, or labellum, acts as a sponge.

▲ MOUTHPARTS
Mosquito
Female mosquitoes have specialized mouthparts that allow them to take blood from animals, or hosts. The mosquitoes need a blood meal to provide nutrients for their eggs.

evolved from hind wings, which were present in the ancestors of flies; like all winged insects, dipterans evolved from four-winged ancestors. Similar structures occur in the strepsipterans or twisted-winged parasites, a small order of bizarre parasitic insects related to the beetles. However, strepsipteran halteres are on the mesothorax; they evolved from the forewings.

The fly's abdomen

The abdomen contains most of the fly's internal organs. Most insect abdomens have 11 segments; muscids such as houseflies, however, have a much reduced number of segments. Only segments 2 to 5 are visible, with segments 6 to 9 greatly narrowed and usually telescoped up inside the rest. These segments form an ovipositor, or egg tube, in females. The reproductive structures of the housefly are mainly internal, but some flies have complex external genitalia for mating.

A tough exoskeleton

All insects are covered by a tough outer covering called an exoskeleton that is secreted from a layer of cells called the epidermis,

which lies directly underneath. The exoskeleton covers the entire outer body surface. It also coats several internal cavities. With the exception of the midgut, the digestive tract is covered by exoskeleton. Most of the internal airways, or tracheae, are covered, too.

The exoskeleton provides support for the body, which does not have an internal skeleton. Inward-pointing struts act as attachment points for an insect's muscles.

COMPARATIVE ANATOMY

Nonflying wings

All true flies have a pair of club-shaped organs called halteres instead of a hind pair of wings. The halteres are formed of three segments. The capitellum is nearest the body; the pedicel is in the middle; and the scabellum is at the tip. The halteres help the fly with balance and stability as it flies with its forewings. By contrast, the front wings of beetles form hardened cases called elytra. They meet along the back of the insect and protect the hind wings when the beetle is resting. The elytra are held straight out to either side of the beetle's body when it is in flight; they may even act as an airfoil to increase lift.

CLOSE-UP

A fly on the wall

Houseflies often land upside down on a ceiling. They are able to do this because their feet are sticky. Each foot has two claws and a pair of cushion-shaped pads called pulvilli. Each pulvillus is covered in tiny tubes that secrete a gluelike substance. A combination of the glue's stickiness plus its surface tension (the elastic nature of its surface film) allows the fly's foot to stick to smooth surfaces, while its claws help it grip tightly to rougher surfaces. The fly detaches the foot by moving it forward, pulling the hairs free from the sticky fluid.

▶ **A fly's foot**
The viscous secretion released by the pulvilli hairs contains a lipid (fat). The secretion forms a very thin layer, and each pulvillus deforms a little to maximize the area of contact.

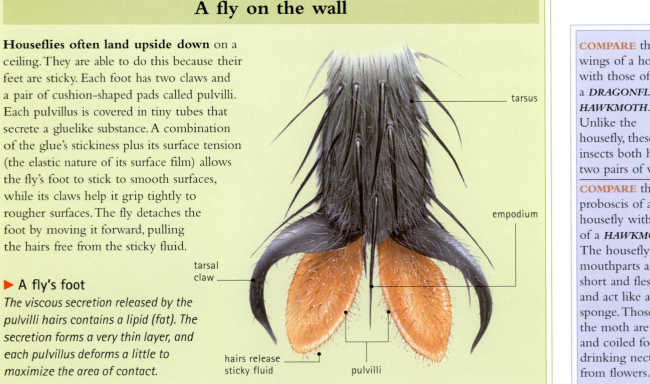

tarsus

empodium

tarsal claw

hairs release sticky fluid

pulvilli

CONNECTIONS

COMPARE the wings of a housefly with those of a **DRAGONFLY** or **HAWKMOTH**. Unlike the housefly, these insects both have two pairs of wings.

COMPARE the proboscis of a housefly with that of a **HAWKMOTH**. The housefly's mouthparts are short and fleshy, and act like a sponge. Those of the moth are long and coiled for drinking nectar from flowers.

GENETICS

Drosophila—the model fly

The vinegar or fruit fly, *Drosophila melanogaster*, is used in laboratories worldwide as a model organism for genetics research. Scientists study fruit flies because they are small, and they breed very quickly. Also, fruit flies have only four pairs of extra-large chromosomes. This is far fewer than most other organisms; humans, for example, have 23 pairs. Chromosomes carry all of the genetic information needed to create a new fly. The information is parceled into units called genes. Genes provide instructions for making proteins, which are assembled in a developing fly embryo to form cells. The cells are organized into tissues, and the tissues are organized into body structures such as an eye or a leg. Sometimes genes mutate (change) and give the wrong instructions. When this happens, mutants result, such as a fly with a leg on its head in place of an antenna or a fly with only one eye. By studying normal and mutant *Drosophila* fruit flies, scientists have come to understand how the process of development from an embryo to an adult works, and which genes give which instructions.

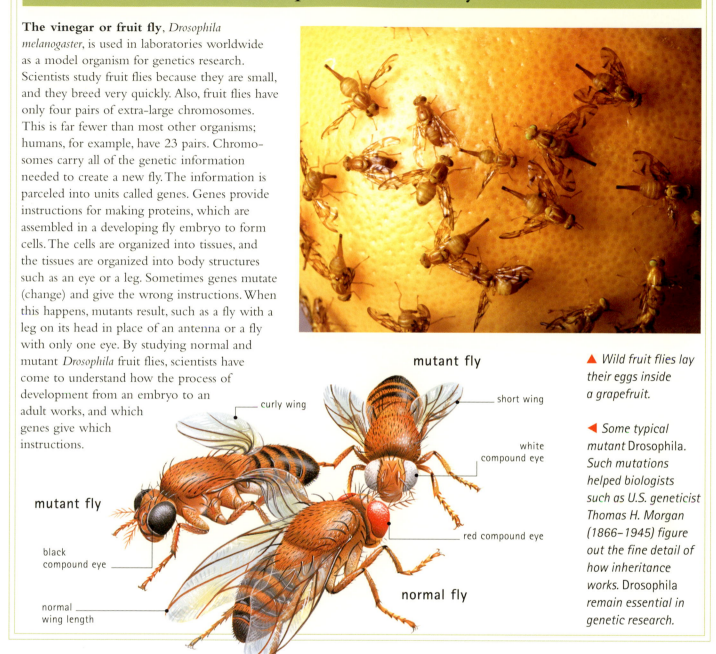

mutant fly

curly wing

mutant fly

black compound eye

normal wing length

short wing

white compound eye

red compound eye

normal fly

▲ *Wild fruit flies lay their eggs inside a grapefruit.*

◄ *Some typical mutant* Drosophila. *Such mutations helped biologists such as U.S. geneticist Thomas H. Morgan (1866–1945) figure out the fine detail of how inheritance works.* Drosophila *remain essential in genetic research.*

In addition, the top layer of the exoskeleton is waterproof and helps minimize water loss from the insect's body. The exoskeleton, together with the evolution of wings, allowed insects to become the dominant animal group on Earth. The exoskeleton enabled early insects to move away from wet environments. They were then able to spread to almost every land habitat, including even the driest deserts.

The exoskeleton is formed mainly of a substance called chitin. This is a nitrogen-rich polysaccharide, or sugar. Chitin is combined in a matrix with various proteins. Most insect body segments are covered with toughened plates of cuticle called sclerites. The cuticle between each two segments is softer to enable the body to move. An elastic protein called resilin makes the cuticle especially flexible where the wings attach to the thorax.

543

Internal anatomy

COMPARE the "open" circulatory system of a housefly with that of **GIRAFFES** or **HUMANS**, which have complex networks of veins and arteries. These animals have a "closed" circulatory system.

CONNECTIONS

Beneath a housefly's exoskeleton are groups of muscles that enable it to move. Some operate the jointed legs to allow the fly to walk; others are responsible for powering the movement of the wings as the animal flies.

There are two sets of muscles in the mesothorax. A vertical set called the dorso-ventral muscles pulls down on the top of the thorax, raising the wings. These muscles then relax and another set, the longitudinal muscles, pulls at the sides of the thorax, raising its upper surface and forcing the wings downward. The downward movement is also aided by the rubbery nature of this part of the exoskeleton, which contains an elastic molecule called resilin. A fly's wing is capable of beating at between 200 and 300 beats per second. This rapid movement causes the buzz you can hear when a fly zooms by.

Hemolymph and hemocoel

A fly's internal organs are surrounded by a cavity called the hemocoel. This contains a bloodlike fluid called hemolymph, which bathes the internal organs. Hemolymph helps remove bacteria and dead cells. It also provides tissues with nutrients and removes carbon dioxide, a waste product of respiration. With

Wing venation

Hemolymph plays an important role in readying the insect for flight. The wings of an adult fly are crumpled and soft when it first emerges from its pupa. Within a few minutes, hemolymph is pumped into the veins, causing the wings to flatten out. Once the wings are fully open, the hemolymph dries and hardens to provide a rigid support structure for each wing membrane.

very few exceptions, insect hemolymph does not contain respiratory pigments and so cannot carry oxygen. Oxygen is instead transported by the tracheal system.

A simple heart running along the top of the body provides some circulation to the hemolymph. Openings from the heart are called ostia. There are accessory pumping (or pulsatile) organs at the base of some of the appendages, such as the wings, legs, and antennae. The pulsatile organs help drive the hemolymph into these structures.

▶ **DIGESTIVE, NERVOUS, AND RESPIRATORY SYSTEMS**
Important features of a housefly's internal anatomy. Flies have an "open" circulatory system, with hemolymph being pushed gently around the body at low pressure by the heart.

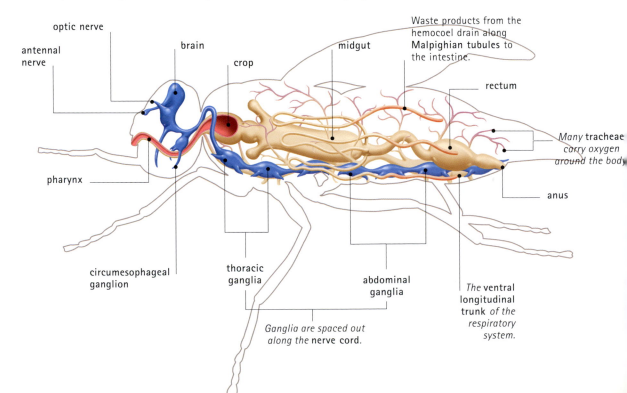

optic nerve
antennal nerve
brain
crop
midgut
Waste products from the hemocoel drain along Malpighian tubules to the intestine.
rectum
Many trachea *carry oxygen around the body*
pharynx
anus
circumesophageal ganglion
thoracic ganglia
abdominal ganglia
The ventral longitudinal trunk *of the respiratory system.*
Ganglia are spaced out along the nerve *cord.*

Nervous system

The central nervous system of an insect consists of pairs of ganglia in each body segment. Ganglia contain the cell bodies of neurons (nerve cells) and are linked by axons (nerve fibers) into a ventral nerve cord. This runs beneath the gut for the length of the body. When an insect receives a stimulus such as a touch to its body, electrical impulses travel from the sense organ (in this case a hair cell) to the nerve cord along sensory neurons. The ganglia process the incoming nerve impulse. They then send out a signal to stimulate an appropriate response from the muscles. Impulses from the ventral nerve cord travel to the muscle along motor neurons.

The brain

The head of a fly is formed from six segments that fused long ago in fly evolutionary history. The ganglia of the first three head segments are fused to form the brain. The brain coordinates sensory information from the eyes, antennae, and labrum (upper mouthpart). The ganglia of the last three head segments are also fused, into the subesophageal ganglion. Lying just below the front part of the gut, this ganglion coordinates the other mouthparts and so controls feeding.

Good fly sight

A housefly has excellent vision. Like most adult insects, it has large compound eyes on either side of its head. Some dipterans are holoptic—their eyes are so large that they meet in the middle of the head. A housefly's compound eye comprises about 4,000 units called ommatidia, each of which forms a facet

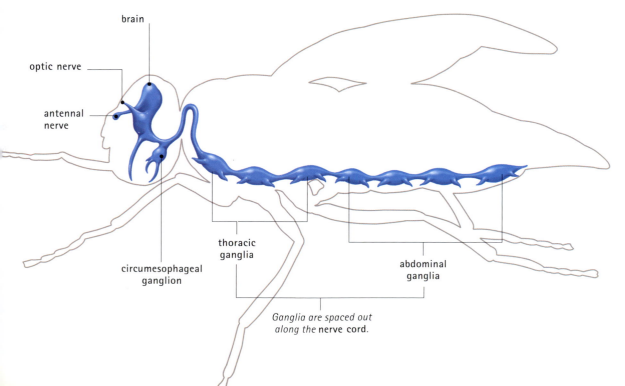

brain

optic nerve

antennal nerve

circumesophageal ganglion

thoracic ganglia

abdominal ganglia

Ganglia are spaced out along the nerve cord.

◀ *Details of a housefly's nervous system. The brain is formed by fused ganglia of the first three head segments. The subesophageal ganglion represents fusion of the next three segmental ganglia.*

545

Taste receptors

Houseflies and other cyclorrhaphan flies can taste their food just by standing on it. Hairs called gustatory sensilla are located on the undersides of their feet. Each hair contains four chemoreceptor cells that are sensitive to different substances, such as sugar or salt. In this way, houseflies are able to distinguish a range of tastes; the hairs help them find appropriate substances to feed on or to lay their eggs in. Unfortunately for humans, though, houseflies also transmit disease-causing bacteria on their feet.

are responsive to odors from other animals, such as cattle or sheep. Novel forms of control for such pests include traps baited with cattle "smell" to which the flies are attracted.

Bloodsucking flies also detect chemicals released by their hosts. Tsetse flies follow trails of carbon dioxide released by practically any vertebrate as they breathe; the tsetses then bite and suck blood from them. Mosquitoes are attracted to chemicals in human sweat or breath. Pregnant women have a slightly raised body temperature, so they release more sweat and they breathe more rapidly. These factors make pregnant women almost twice as likely as other people to be bitten by mosquitoes. They stand a higher risk of contracting mosquito-borne diseases such as malaria.

▼ *The head of a horse fly. Note the individual lenses of the compound eyes. The short antennae contain cells that detect airborne chemicals. Cells connected to hairs at the base of the labellum (at the bottom of the mouthparts) work in tandem with cells on the feet to taste food.*

of the eye. The fly sees the world as a mosaic picture formed from the individual images from each ommatidium. The eyes are particularly attuned to movement—this explains why houseflies are so adept at avoiding flyswatters.

In addition to the compound eyes, flies have three smaller, simple eyes, called ocelli, on the top of the head. They are particularly sensitive to changes in light intensity.

How flies touch

A housefly's body is covered with many hairs and bristles, all of which are connected to sensory neurons and provide information on touch. These sensitive hairs are called mechanoreceptors. Most insects also use their antennae for sensing touch. Housefly antennae are short and have few segments compared with those of some other types of flies. These segments consist of a scape that attaches to the head; this links to the pedicel, which attaches in turn to the bean-shaped annulus. From the annulus extends a sensitive, featherlike structure called the arista.

Smelling chemicals

The antennae bear sensory hairs called olfactory sensilla. They respond to airborne chemicals, such as pheromones. Male insects are often sensitive to pheromones released by females as a signal to mate. Many dipterans, such as those that lay their eggs on livestock,

Circulatory and respiratory systems

Tubes called tracheae permit the uptake of oxygen from the air into the fly's body tissues. They open through pores called spiracles that lie on each side of the abdomen and thorax. The tracheae are actually ingrowths of the exoskeleton (the hard outer skin of the animal) and are lined with cuticle.

In houseflies the tracheae widen into air sacs, but eventually they divide into tiny tubes called tracheoles. Their tips are fluid-filled. There, oxygen diffuses (moves) from the air into the tissues. Oxygen is needed for cellular respiration, the process by which cells derive energy from glucose sugar. A waste product of respiration, carbon dioxide, diffuses in the opposite direction to leave the body through the spiracles. The opening and closing of the spiracles is triggered by carbon dioxide levels. They are kept closed for as long as possible to minimize water loss.

In and out

Many insects do not actively breathe in and out as animals with lungs do. Instead, they mainly rely on the passive diffusion of gases. When an insect is active, it can draw fluid out of the tracheoles to leave more room for air. Hover flies, and some other insects such as wasps, ventilate their tracheae by pumping air in and out, using collapsible air sacs.

Many fly larvae live in environments where oxygen is scarce. Rat-tailed maggots are hover fly larvae that live at the bottom of pools and puddles. They breathe through long telescopic tubes, which they poke above the water's surface. Mosquito and midge larvae also live in stagnant, oxygen-poor water. Mosquito larvae have breathing tubes through which they obtain oxygen as they hang beneath the water's surface. Midge larvae are known as bloodworms because they are bright red with hemoglobin. This is a respiratory pigment that also occurs in the blood of vertebrates. It allows a midge larva to absorb and retain the tiny amount of oxygen present in the water.

Respiratory pigments are rare in insects; however, another insect that uses hemoglobin is the horse botfly larva. This larva develops inside the stomach of horses. Other parasitic flies retain a connection to the outside. For example, when warble fly maggots burrow into the flesh of their cattle hosts, they leave a large breathing hole at the surface. This gives their spiracles access to the air.

Housefly larvae have two large black spiracles at the rear of their body. These openings remain exposed to the air to allow the larva to breathe as it feeds head down in its food, which may be an anaerobic environment (one lacking in oxygen).

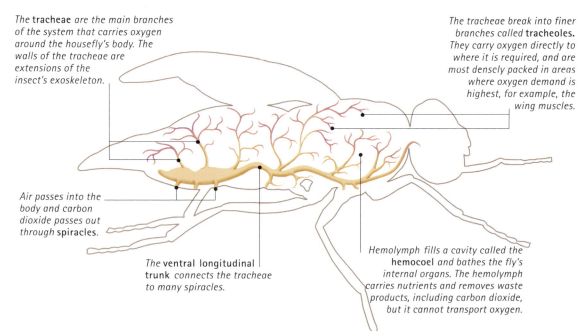

The **tracheae** are the main branches of the system that carries oxygen around the housefly's body. The walls of the tracheae are extensions of the insect's exoskeleton.

The tracheae break into finer branches called **tracheoles.** They carry oxygen directly to where it is required, and are most densely packed in areas where oxygen demand is highest, for example, the wing muscles.

Air passes into the body and carbon dioxide passes out through **spiracles**.

The **ventral longitudinal trunk** connects the tracheae to many spiracles.

Hemolymph fills a cavity called the **hemocoel** and bathes the fly's internal organs. The hemolymph carries nutrients and removes waste products, including carbon dioxide, but it cannot transport oxygen.

◄ Tracheae are lined by extensions of the exoskeleton. This means that in young insects, the lining of the whole system must be shed with the rest of the skin to allow the insect to grow. Air sacs, which some insects use to flush the tracheal system, are not covered by cuticle. This allows them to be compressed. The hemolymph carries carbon dioxide—but not oxygen.

547

Digestive and excretory systems

The housefly's digestive tract is a simple tube stretching from mouth to anus. It functions by digesting food and absorbing nutrients and water. Excretion of the nitrogen-containing waste products of the body's metabolism occurs through a series of tubes called the Malpighian tubules, which filter the fly's hemolymph. Hemolymph is the fluid that circulates around the body cavity, or hemocoel.

External digesters

Before houseflies can take in food, they must liquefy it by external digestion. To do this they regurgitate down the hypopharynx part of a previous meal plus some saliva, and dribble the mixture onto the food. Enzymes from the midgut and saliva begin to digest the food, breaking it down so that it can be sucked up by the spongelike mouthparts.

From gut to tissues

A muscular pharynx in the foregut helps draw digested food into the digestive tract. The foregut and hindgut of insects are lined with cuticle. The midgut has a membrane across which digested food is transferred into the hemolymph. The hemolymph then carries the nutrients to the cells that need them. The

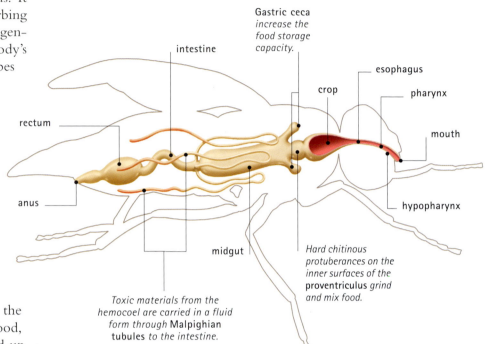

Gastric ceca *increase the food storage capacity.*

intestine

esophagus

crop

pharynx

mouth

rectum

anus

hypopharynx

midgut

Hard chitinous protuberances on the inner surfaces of the **proventriculus** *grind and mix food.*

Toxic materials from the hemocoel are carried in a fluid form through **Malpighian tubules** to the intestine.

▲ *A fly's digestive system. Food is partly digested outside the body. The pharynx or cibarial pump then sucks the soup of broken-down food through the mouthparts.*

storage capacity of the gut is increased by a series of pouches, called ceca. Almost all adult flies feed on foods different from those that their larvae consume. Most adults feed on nectar from flowers or other such sweet liquids. Many female flies, such as mosquitoes, midges, and tsetse flies, require a blood meal before they lay eggs. Others, such as dung flies, prey on other insects, and some adult flies may not eat at all.

The excretory system

The removal of metabolic waste, or excretion, occurs through the Malpighian tubules, which are located near the junction of mid- and hindgut. These blind-ended tubes float freely in the hemolymph. Nitrogenous waste is drawn into the lumen (cavity) of the tubules. The waste then passes into the hindgut.

Water is reabsorbed from excretory waste in the rectum at the end of the digestive tract. The resulting urine is highly concentrated; it passes from the digestive tract through the anus along with undigested material from the midgut.

IN FOCUS

Larval digestion

Most female flies lay their eggs directly onto the food that their larvae will eat. In houseflies, this might be rotting food in a trash can or a compost heap, but in other flies it may be dung, rotting meat, or even living flesh. Some flesh-eating larvae cause serious damage to the health of domestic animals. Maggots of the sheep blow fly cause a condition called "sheep strike," or myiasis, which affects millions of sheep each year. Surprisingly, the same larvae have medicinal uses; they can be used to clean human wounds by eating away the dead flesh.

Reproductive system

All true flies reproduce sexually; that is, sperm from a male fertilizes the eggs of a female. Male and female houseflies each have one pair of gonads located in the abdomen. In females, two ovaries pass eggs into a common oviduct. The oviduct, accessory glands, and a vessel for storing sperm called the spermatheca all open into a structure called the vagina. In males, two testes produce sperm. This is stored in the seminal vesicles, which lie on either side of an ejaculatory duct.

Egg-laying and live-bearing

Not all flies are oviparous (egg-laying); some give birth to live offspring. The sheep botfly is ovoviviparous. The eggs hatch inside the female's body, and she lays her larvae directly inside the nostrils of domestic sheep. When they are ready to pupate, the maggots are sneezed out onto the ground. Female tsetse flies retain their larvae for even longer; they are referred to as viviparous (live-bearing). A single larva feeds on a secretion from glands inside the female; after developing for a time it is deposited onto the soil, into which it burrows before pupating.

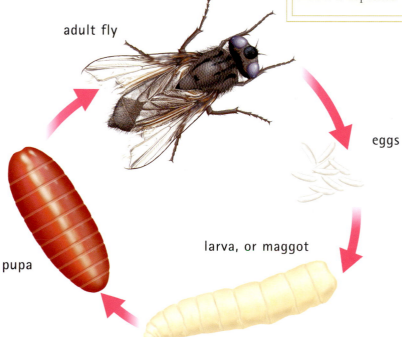

adult fly

eggs

larva, or maggot

pupa

CLOSE-UP

A complete change

Like beetles, wasps, butterflies, and a few other groups, houseflies undergo complete metamorphosis during their life cycle. Adults of these insects look remarkably different from the juveniles. They live in different places and exploit different food sources. The larvae spend most of their time feeding. By contrast, the adults spend most of their time moving around and reproducing.

▲ LIFE CYCLE
Houseflies pass from the larval to adult stage through an inactive pupal stage. They burst from the pupa by using an inflatable sac called the ptilinum.

Laying eggs and growing up

Fertilization takes place internally following copulation. The accessory glands play a crucial role; their secretions break down the outer membranes of the sperm and part of the egg, so fertilization can take place. Eggs are laid through an ovipositor, which is stored telescopically inside the abdomen but pops out when egg-laying takes place. Eggs are deposited inside a good food source for the young, such as decomposing trash. A female housefly may mate only once but, because sperm is stored in the spermatheca, she may lay several batches of eggs during her lifetime.

Female houseflies lay up to 250 eggs. The eggs are white and around 0.04 inches (0.1 cm) long. Larvae called maggots hatch from them. The maggots grow rapidly to a length of around 0.5 inch (1.2 cm) before pupating. During pupation, the larva secretes a tough case called a puparium. This barrel-shaped structure is 0.2 inch (0.6 cm) long and changes from yellow to black during pupation. Inside, the fly undergoes a major reorganization of its structure, a process called metamorphosis.

KATIE PARSONS

FURTHER READING AND RESEARCH
Blum, Mark, 1998. *Bugs in 3-D*. Chronicle Books: San Francisco, CA.
2003. *Insects and Spiders of the World*. Marshall Cavendish Corporation: Tarrytown, NY.

Human

ORDER: **Primates** SUBORDER: **Catarrhini** FAMILY: **Hominidae**
GENUS: *Homo*

The modern human, *Homo sapiens*, is the only surviving member of the genus *Homo*. There have been several other species in the genus *Homo*; they walked upright and had a relatively large brain. The earliest member of the genus, *Homo habilis*, evolved from another hominid (humanlike ancestor), *Australopithecus*, around 2.4 million years ago. Modern humans did not appear until around 100,000 years ago. They shared the world with one or more other types of humans, including the Neanderthal people of Europe and western Asia.

Anatomy and taxonomy

Human classification is riddled with controversy. Anatomical studies have been relatively ineffective for determining human relationships with other hominids, but genetic research has proved a valuable tool.

● **Animals**   All animals are multicellular. They get the energy and materials they need to survive by consuming other organisms. Unlike other multicellular organisms such as plants and fungi, animals are able to move around for at least one phase of their lives.

● **Chordates**   At some time in their life cycle, all chordates have a stiff supporting rod called a notochord running along the back of their body.

● **Vertebrates**   In vertebrates the notochord changes into a backbone made up of units called vertebrae. The spinal cord runs through the backbone. Most vertebrates are bilaterally symmetrical; the body shape is roughly the same on each side of the backbone. Surrounding the brain, all vertebrates have a skull made of either bone or cartilage.

● **Mammals**   Mammals have hair, a single lower jawbone that hinges directly onto the skull, and red blood cells that lack nuclei. Female mammals provide their young with milk. Mammals are able to create their own body heat (they are warm-blooded) and have a four-chamber heart.

● **Placental mammals** A placental mammal develops inside its mother's uterus, where it receives nourishment and oxygen from the mother through an organ called the placenta. This develops during pregnancy. The other two

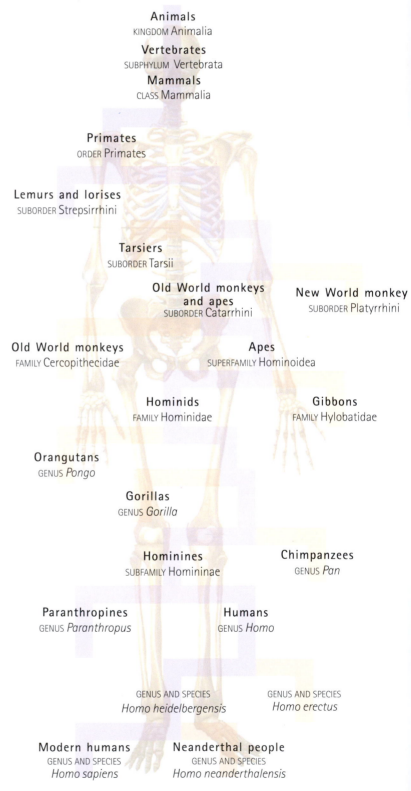

▼ *The taxonomic relationships shown below represent the consensus of many biologists but are by no means universally accepted. Human taxonomy is fraught with difficulty and is constantly changing. While some biologists consider* Homo heidelbergensis, Homo erectus, *and* Homo neanderthalensis *to be separate species, others believe the last to be a subspecies of* Homo sapiens.

Animals
KINGDOM Animalia

Vertebrates
SUBPHYLUM Vertebrata

Mammals
CLASS Mammalia

Primates
ORDER Primates

Lemurs and lorises
SUBORDER Strepsirrhini

Tarsiers
SUBORDER Tarsii

Old World monkeys and apes
SUBORDER Catarrhini

New World monkey
SUBORDER Platyrrhini

Old World monkeys
FAMILY Cercopithecidae

Apes
SUPERFAMILY Hominoidea

Hominids
FAMILY Hominidae

Gibbons
FAMILY Hylobatidae

Orangutans
GENUS *Pongo*

Gorillas
GENUS *Gorilla*

Hominines
SUBFAMILY Homininae

Chimpanzees
GENUS *Pan*

Paranthropines
GENUS *Paranthropus*

Humans
GENUS *Homo*

GENUS AND SPECIES
Homo heidelbergensis

GENUS AND SPECIES
Homo erectus

Modern humans
GENUS AND SPECIES
Homo sapiens

Neanderthal people
GENUS AND SPECIES
Homo neanderthalensis

mammal groups are the marsupials, which give birth to tiny young that complete their development in the mother's pouch; and monotremes, which lay eggs.

● **Primates** Primates form a large group of mammals that include prosimians, such as bush babies and lemurs, monkeys, and apes. There are around 279 species of primates; they have a well-developed cerebral hemisphere enclosed by a large globe-shaped cranium. Most primates have short jaws and flat faces, with short noses and large, forward-pointing eyes. For most primates the sense of smell is less important than vision, touch, and hearing.

● **Hominids** The family Hominidae traditionally consisted solely of humans and their immediate ancestors and relatives, but recently the great apes (the chimpanzees, gorillas, and orangutans) have been included. Hominids have a short spine with shoulder blades that provide an exceptionally wide range of movement for the arms.

▲ *Unlike many other mammals, humans have relatively little facial hair. Facial expressions are therefore easy to read.*

● **Hominines** This group includes all the non-ape hominids. Humans are the sole living representatives, though there are many fossil species. These include *Australopithecus*, the group from which true humans evolved. Hominines are bipedal. Unlike those of the great apes, the first two digits on humans' feet are not opposable.

● **Humans and very close relatives** The genus *Homo* includes modern humans, plus a number of extinct relatives such as Neanderthals and *Homo floresiensis*, a species of 3-foot (90-cm) people. *Homo* differs from earlier hominines such as *Australopithecus* in having a skeleton better adapted to standing upright, slighter jaws and chewing muscles, and a particularly large brain. Scientists may soon agree to re-classify hominids on the basis of new evidence. The new classification would place chimpanzees in the genus *Homo*.

FEATURED SYSTEMS

EXTERNAL ANATOMY Humans are bipedal with an upright posture, leaving the hands free to manipulate objects. *See pages 552–554.*

SKELETAL SYSTEM The human skeleton shows unique adaptations to cope with a massively increased brain, a softer diet, and an upright, bipedal stance. *See pages 555–558.*

MUSCULAR SYSTEM Some muscles are well-developed for maintaining an upright position and a well-balanced bipedal gait. *See pages 559–561.*

NERVOUS SYSTEM Humans have the most complex central nervous system in the animal kingdom, and have unique problem-solving and cognitive abilities. *See pages 562–565.*

CIRCULATORY AND RESPIRATORY SYSTEMS Humans have a four-chamber heart and a pair of lungs inflated by the action of the diaphragm. *See pages 566–567.*

DIGESTIVE AND EXCRETORY SYSTEMS Humans have a relatively simple digestive system with few unique adaptations. *See pages 568–569.*

ENDOCRINE AND EXOCRINE SYSTEMS These systems are groups of glands. One type, mammary glands, occurs only in mammals. *See pages 570–571.*

REPRODUCTIVE SYSTEM The female reproductive system allows for the efficient development of one or two offspring at a time. As in most mammals, the male's testes are held outside the body. *See pages 572–573.*

External anatomy

CONNECTIONS

COMPARE humans' upright stance with the stances of other bipeds, such as **KANGAROOS** or **WOODPECKERS**.

COMPARE human feet with those of a **CHIMPANZEE**.

Biologically, humans are among the most unusual mammals. For example, humans have an erect posture and are bipedal. Almost all other mammals are quadrupedal, or four-legged. The only other truly bipedal mammals are kangaroos, and some rodents such as jerboas. Although monkeys and apes occasionally walk upright, they usually move around on four legs.

Why walk on two legs?

Bipedalism leaves the front limbs free to carry and use tools; thus it may have kick-started humans' rapid increase in brain size. Despite its

▲ Humans are able to balance on two legs with the aid of organs that form part of the inner ear.

importance, however, biologists have little idea why human ancestors began to walk on two legs rather than four. Bipedality may have evolved after a change in habitat; climatic change around 4 million years ago brought a decline in tropical forests, so human ancestors were forced to switch to life on grasslands.

Biologists often suggest that bipedalism may have been advantageous for observing predators, or for reducing the surface area of body exposed to the hot sun. However, other grassland animals, such as baboons, do just fine walking on four legs. Perhaps the likeliest explanation involves the energetics of walking. Bipedal movement is more efficient than a

Forward-facing **eyes** provide humans with stereoscopic vision. This enables people to accurately judge distances.

The head contains the most important sense organs: the **eyes**, **nose**, **tongue**, *and* **ears**.

Like all adult female mammals, human females have **mammary glands**, which provide nutrient-rich milk for newborn young.

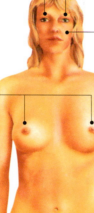

Human **thumbs** *are opposable*; they can be placed against the other digits. This ability enables humans to handle objects with great dexterity.

Human **genitals** *are similar to those of other mammals; the female has a* **vagina** *and the male a* **penis** *and* **testicles**.

► **Female human**

Humans have very little body hair in comparison with most other mammals. They are also sexually dimorphic: males and females look noticeably different. Females have a broader pelvis, breasts, and less body hair than males.

Walking on two **legs**— bipedality—is an unusual trait shared by only a handful of other mammals such as kangaroos and jerboas.

Humans are **plantigrade**; they walk on the soles of their feet rather than on their toes.

Unlike other primates, humans do not have opposable **toes**.

The whites of the eyes

Humans are one of the few mammals in which the whites of the eyes are visible. This feature makes it easier to determine where a person is looking and might help humans communicate. The eye whites are visible in some other social animals such as wolves.

chimp's quadrupedal knuckle-walking. For a knuckle-walking ancestor living on the plains, switching to bipedality would have offered the best means of moving long distances in the absence of trees to swing from.

Limb dimensions

Unlike other apes, humans have long, powerful legs and relatively short arms. Humans are plantigrade; that is, they walk on the soles of the feet with the heels on the ground. The arm span of some apes can be more than twice their height, whereas arm span and height are almost equal in humans. Although humans do not use their arms for locomotion, the arms are relatively powerful, reflecting their former use for swinging from branch to branch and their importance for tool manipulation. As in all apes, human hands have five fingers including an opposable thumb. It is the opposable thumb, an adaptation for grasping branches when climbing, that has allowed humans to make and manipulate tools and change the environment in which they live.

▼ SKIN CROSS SECTION

This diagram shows the structures typically found in the epidermis (outer skin layer), the dermis (middle layer), and a portion of the thickest layer, the hypodermis (lower layer), of human skin. The outer layer of skin cells is constantly shed and replaced by new cells rising from the lower epidermis.

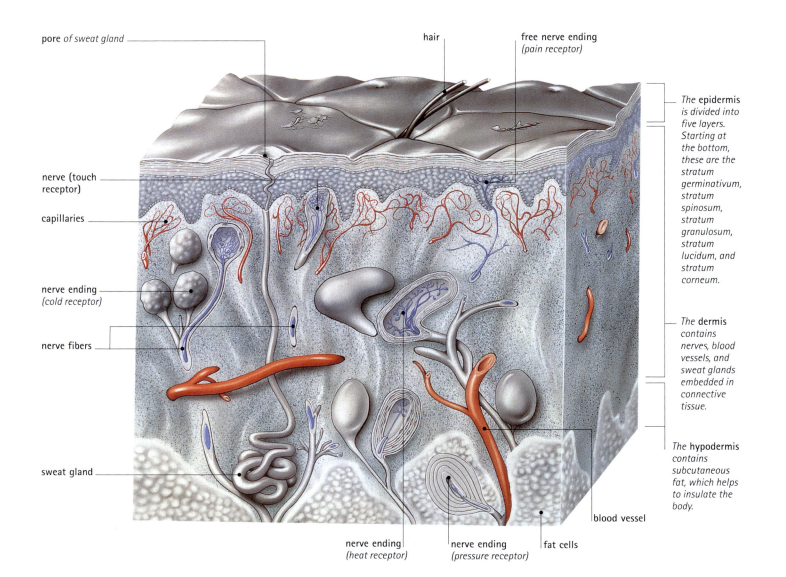

pore *of sweat gland*

hair

free nerve ending *(pain receptor)*

nerve (touch receptor)

capillaries

nerve ending *(cold receptor)*

nerve fibers

sweat gland

nerve ending *(heat receptor)*

nerve ending *(pressure receptor)*

fat cells

blood vessel

The **epidermis** *is divided into five layers. Starting at the bottom, these are the stratum germinativum, stratum spinosum, stratum granulosum, stratum lucidum, and stratum corneum.*

The **dermis** *contains nerves, blood vessels, and sweat glands embedded in connective tissue.*

The **hypodermis** *contains subcutaneous fat, which helps to insulate the body.*

▶ *The beard worn by this Confucian holy man has become white with age. A hair grows from a group of cells under the skin called a follicle. As people grow older, the cells in the follicle that produce pigment gradually die and the hairs grow white.*

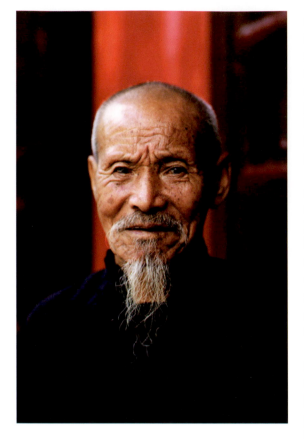

Where did all the hair go?

Another biological enigma is humans' relative lack of hair. Hairs are present all over the body (apart from on the lips, palms, and soles) in numbers similar to those of other primates. However, human hairs are short and very fine. Biologists struggle to explain this unusual feature. The reduction in hair occurred not less than 70,000 years ago, when people began to wear clothes, but may have evolved long before that. Biologists have suggested a variety of explanations for the reduction in hair: an aquatic human ancestry; a means of foiling parasitic insects; and a way of keeping cool in hot climates. None of these ideas stands up to close scrutiny, however. The most convincing theory suggests that hair reduction evolved for sexual display, perhaps to advertize the absence of parasites and therefore suitability for mating.

Explaining the anomalies

How can the unusual distribution of the remaining human hair be explained? Women have hair on the top, sides, and back of the head, leaving the face and forehead only sparsely covered with tiny fine hairs, apart from conspicuous eyebrows. Most adult men have a dense covering of hair on their head and also on the lower face. This facial hair starts to develop in human males when they reach puberty at about 13 years old. Hair on top of the head probably helps protect the brain from the heat of the sun. The hairs of the eyebrows channel sweat and water away from the eyes.

A partially or wholly naked face may have allowed human ancestors to communicate visually with a wide range of facial expressions. It may also have indicated a person's level of health (and therefore sexual desirability). What about beards? Their presence in men but not women strongly suggests that evolution favored the retention of beards as an ornament of sexual display and status, either for display to other males or to attract females.

Other external features

Compared with other primates, humans have a higher forehead and smaller jaws that do not project as far forward. Modern humans have a larger, more spherical head, accommodating a larger brain than that of their hominid ancestors, and they have smaller jaws.

Like other primates, humans have forward-facing eyes and good stereoscopic (three-dimensional) vision. This feature was important for human ancestors' tree-living lifestyle, when it was necessary to judge distances accurately while jumping from bough to branch. It also enables detailed close-up vision.

CLOSE-UP

Explaining pubic hair

As well as on the head, the otherwise hairless human has retained hair in the armpits and pubic regions. In human ancestors, these hairs may have served as a large surface from which to broadcast chemicals called pheromones into the air. These chemicals served as a sexual attractant. Modern humans may still release attractant pheromones, although this possibility has been little studied by biologists.

Skeletal system

At birth, a human's skeleton is made up of 270 bones. During development from baby to adult many of the bones fuse, so by the age of about 20 the total number of bones has reduced to 206. The human skeleton can be divided into two main sections: the axial skeleton consists of the skull, vertebral column, and ribs; and the appendicular skeleton is made up of the limbs and limb girdles. There are three main types of bones in the human skeleton: long bones, flat bones, and short bones. Long bones, such as leg and arm bones, are long and tubular, usually with joints for movement at each end. Flat bones have a flat cross section, and include the ribs and shoulder blades. Short bones have irregular shapes and include the vertebrae and the small block-shaped bones in the hands and feet.

Skeletal functions

The skeleton provides support and structure for the whole body. The legs and back support the weight of the rest of the body. The skeleton allows movement by providing attachments for muscles to pull against. It also protects internal organs such as the heart, lungs, and brain. The marrow of the long bones is the source of white and red blood

CLOSE-UP

The smallest bones

The smallest bones in the human body are the ossicles, a series of three tiny bones located in the middle ear. These bones are named the malleus, the incus, and the stapes. They transfer sound vibrations from the eardrum to the inner ear, where they are converted to nerve signals sent to the brain. The ossicles are unique to mammals. Reptiles and amphibians have a stapes, but the mammalian incus and malleus evolved from some of the bones that formed the lower jaw in mammals' reptilian ancestors.

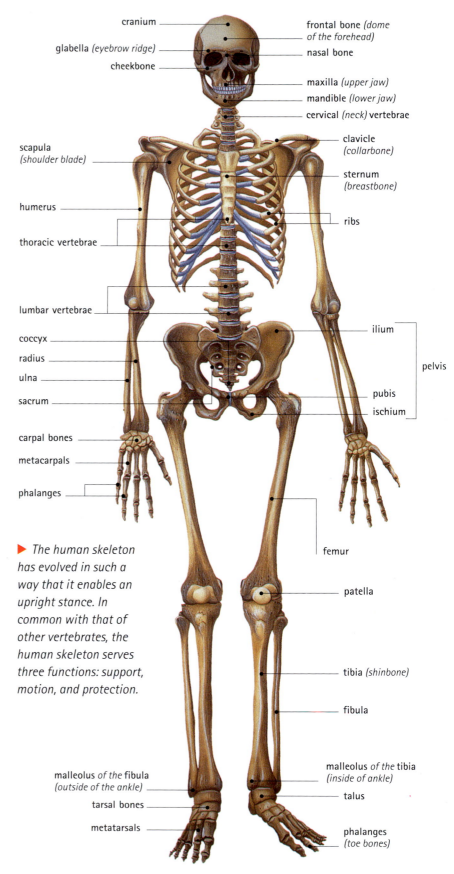

cranium — frontal bone (dome of the forehead)
glabella (eyebrow ridge) — nasal bone
cheekbone — maxilla (upper jaw)
mandible (lower jaw)
cervical (neck) vertebrae
scapula (shoulder blade) — clavicle (collarbone)
sternum (breastbone)
humerus — ribs
thoracic vertebrae
lumbar vertebrae
coccyx — ilium
radius
ulna — pelvis
sacrum — pubis
ischium
carpal bones
metacarpals
phalanges
femur
patella
tibia (shinbone)
fibula
malleolus of the fibula (outside of the ankle) — malleolus of the tibia (inside of ankle)
talus
tarsal bones
metatarsals — phalanges (toe bones)

▶ The human skeleton has evolved in such a way that it enables an upright stance. In common with that of other vertebrates, the human skeleton serves three functions: support, motion, and protection.

Shifts of the foramen magnum

The foramen magnum is an opening at the base of the skull where nerves of the medulla oblongata connect the brain to the spinal cord. The position of the foramen magnum in ancient hominids gives a clue to how they moved. In hominids, unlike chimpanzees and other apes, the foramen magnum opens toward the front of the skull; its forward position is particularly pronounced in modern humans. This is the best arrangement for bipedality, since it allows the head to sit directly on top of the shoulders. The foramen magnum of a quadrupedal animal is located farther back toward the rear of the skull.

cells, crucial for oxygen transport and for immunity against diseases. Bones also provide a store of essential minerals such as calcium and phosphorous.

The skull

The human skull is made up of at least 22 bones, eight of them in the top, back, and sides of the head (the cranial bones), and the rest in the facial area. In newborn babies, many of these bones are only loosely joined by fibrous elastic tissue; that allows the baby's large head easier passage through the mother's birth canal. Later, these joints, or sutures, in the skull fuse, providing a protective helmet for the brain.

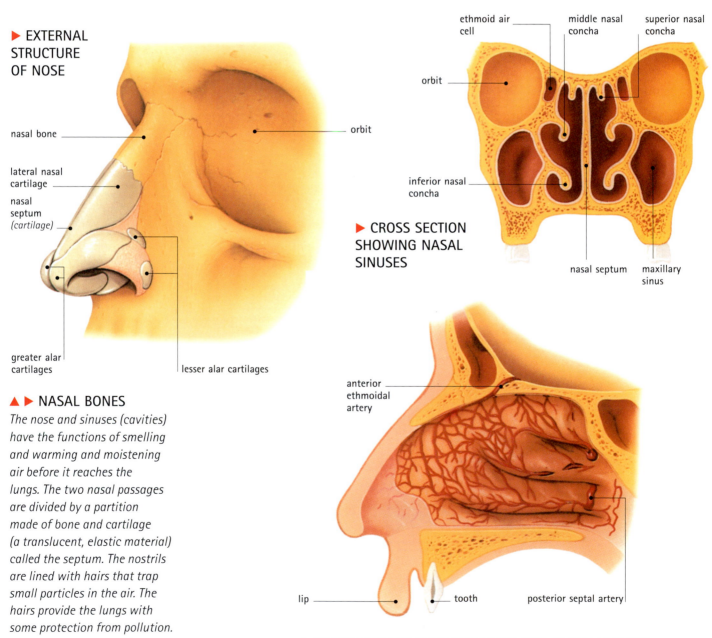

► EXTERNAL STRUCTURE OF NOSE

nasal bone

orbit

lateral nasal cartilage

nasal septum (cartilage)

greater alar cartilages

lesser alar cartilages

► CROSS SECTION SHOWING NASAL SINUSES

ethmoid air cell

middle nasal concha

superior nasal concha

orbit

inferior nasal concha

nasal septum

maxillary sinus

anterior ethmoidal artery

lip

tooth

posterior septal artery

▲► NASAL BONES

The nose and sinuses (cavities) have the functions of smelling and warming and moistening air before it reaches the lungs. The two nasal passages are divided by a partition made of bone and cartilage (a translucent, elastic material) called the septum. The nostrils are lined with hairs that trap small particles in the air. The hairs provide the lungs with some protection from pollution.

▲ CROSS SECTION SHOWING BONE AND BLOOD VESSELS

Ancient diets

Skull shape can tell scientists a lot about the diets of extinct hominids. Early hominids such as *Paranthropus boisei* had large jaws and enlarged bony processes on the skull for the attachment of powerful chewing muscles. These structures, and the presence of large molar teeth, reveal that these creatures ate tough fibrous food that needed a lot of chewing. Other hominids such as *Homo habilis* had much smaller jaws and teeth. *Homo habilis* was omnivorous, feeding on plant and animal matter; it used tools to help it cut animal hide or break up bones to get at the marrow within. Even though modern humans have smaller jaws and jaw muscles than our predecessors, a human bite is still powerful and can inflict injuries comparable to a dog bite.

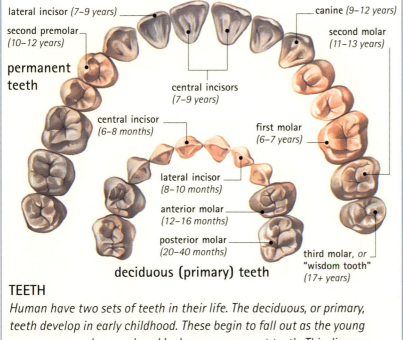

lateral incisor *(7–9 years)*

second premolar *(10–12 years)*

permanent teeth

central incisors *(7–9 years)*

central incisor *(6–8 months)*

lateral incisor *(8–10 months)*

anterior molar *(12–16 months)*

posterior molar *(20–40 months)*

deciduous (primary) teeth

canine *(9–12 years)*

second molar *(11–13 years)*

first molar *(6–7 years)*

third molar, *or* "wisdom tooth" *(17+ years)*

TEETH
Human have two sets of teeth in their life. The deciduous, or primary, teeth develop in early childhood. These begin to fall out as the young person ages and are replaced by larger permanent teeth. This diagram shows the approximate ages at which different teeth first appear.

As well as housing the brain, the skull also holds the most important human sense organs: the eyes, nose, ears, and tongue.

The human skull has a number of holes, or foramina, particularly around the base. These holes allow various blood vessels and nerves to pass through, going to or from the brain. The largest of these holes is the foramen magnum, through which pass various important arteries and the medulla oblongata, the lower part of the brain stem that continues downward to form the spinal cord.

The vertebral column and ribs

Humans have 33 vertebrae arranged in five groups. Flexible fibrous disks between the vertebrae allow bending movement in the spine and act as shock absorbers against the jolting forces that occur during running or jumping. The uppermost seven vertebrae are the cervical vertebrae; they give the neck flexibility. The top two of these bones are the atlas, which allows the head to nod up and down, and the axis, which allows the head to move from side to side.

Next come the 12 thoracic vertebrae, to which the rib cage attaches. The first seven

▼ *The skeleton provides support for movements such as bending and lifting. Muscles attached to the bones provide mechanical force, while the bones provide a firm structure against which the muscles can move.*

IN FOCUS

Skeletal changes for bipedality

The structure and arrangement of many human skeletal features enable bipedality. This is especially true of the pelvic girdle and legs. The human pelvis is shorter and wider than that of a chimp; also, the socket joints that accommodate the heads of the femurs (thighbones) point downward, so the weight of the body passes efficiently through the hip. The separated hipbones, or ilia, allow a wide stance, important for good balance on two legs. The hipbones also act as an attachment point for the gluteal muscles, which aid balance. The femurs point inward a little from the hips to the knees, making it easier for humans to remain balanced when shifting their weight from one foot to the other during walking. Human knee joints are broad to help keep a stable posture. The lack of an opposable toe and the development of a large heel bone (or talus) have made human feet efficient for walking.

FEMALE PELVIS   MALE PELVIS

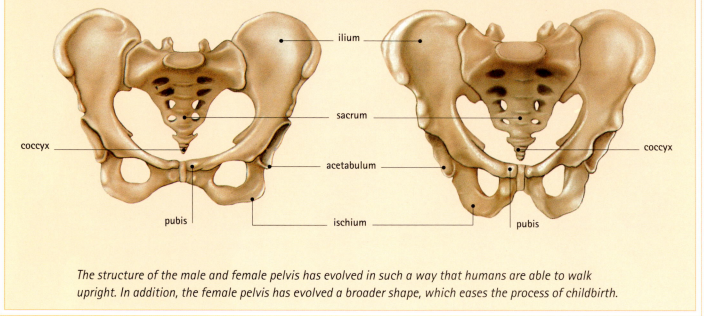

ilium

sacrum

coccyx

acetabulum

coccyx

pubis

ischium

pubis

The structure of the male and female pelvis has evolved in such a way that humans are able to walk upright. In addition, the female pelvis has evolved a broader shape, which eases the process of childbirth.

pairs of ribs connect via cartilage to a long, flat bone at the front of the chest called the breastbone, or sternum. The sternum is also joined at its top to the collarbones, or clavicles. The ribs form a cage that protects delicate internal organs such as the heart and lungs.

The next five vertebrae are the largest of the vertebral bones and support most of the body's weight. These are the lumbar vertebrae, which support the lower back. Below the lumbar region is the sacrum, a series of four or five sacral vertebrae that become fused in adults to provide sturdy support for the pelvis.

The final section of the vertebral column is the coccyx, or tailbone, which is made up of three to five vertebrae that fuse by adulthood. The coccyx curves under the pelvis and serves as a site for muscle attachment.

The limbs

There is a total of 126 bones in the human appendicular skeleton, which includes the bones of the arms and legs, the pelvis, and the shoulder area. Like apes, humans have five-fingered hands with opposable thumbs. In humans and apes, a series of long finger bones, or phalanges, give the hands considerable dexterity and the ability to grasp objects. The upper arm bone, or humerus, is able to rotate freely in the shoulder joint, an adaptation that allows gibbons, apes, and humans' distant ancestors to move through trees by swinging from one hold to another.

Human leg bones, such as the femur (thighbone) and the fibia and tibia (shinbones), are much longer than those of apes. The femur is the largest bone in the human body.

Muscular system

In common with all vertebrates, humans have three kinds of muscles: cardiac, or heart, muscle; smooth muscle, which is mostly involuntary (it operates without conscious thought); and skeletal muscle, which is under conscious control. Skeletal muscles enable humans to move around and account for around 40 percent of body weight in adult males, but only 23 percent in females. There are more than 600 skeletal muscles in the human body. Muscles of all shapes and sizes are made up of blocks of muscle fibers. These fibers contract when they receive signals from motor neurons (nerve cells). Since muscles can exert only a pulling force, they often work in antagonistic pairs, able to pull against each

▼ **Muscles**
The main voluntary muscles on the front and back of a human. The voluntary muscles are those that are under conscious control and enable humans to move.

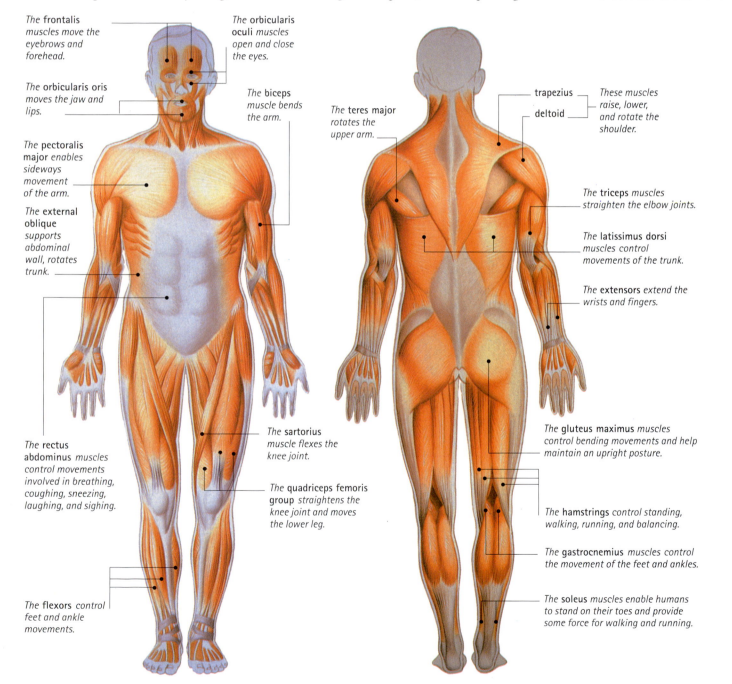

The **frontalis** *muscles move the eyebrows and forehead.*

The **orbicularis oculi** *muscles open and close the eyes.*

The **orbicularis oris** *moves the jaw and lips.*

The **biceps** *muscle bends the arm.*

The **pectoralis major** *enables sideways movement of the arm.*

The **external oblique** *supports abdominal wall, rotates trunk.*

The **rectus abdominus** *muscles control movements involved in breathing, coughing, sneezing, laughing, and sighing.*

The **sartorius** *muscle flexes the knee joint.*

The **quadriceps femoris group** *straightens the knee joint and moves the lower leg.*

The **flexors** *control feet and ankle movements.*

The **teres major** *rotates the upper arm.*

trapezius

deltoid

These muscles raise, lower, and rotate the shoulder.

The **triceps** *muscles straighten the elbow joints.*

The **latissimus dorsi** *muscles control movements of the trunk.*

The **extensors** *extend the wrists and fingers.*

The **gluteus maximus** *muscles control bending movements and help maintain an upright posture.*

The **hamstrings** *control standing, walking, running, and balancing.*

The **gastrocnemius** *muscles control the movement of the feet and ankles.*

The **soleus** *muscles enable humans to stand on their toes and provide some force for walking and running.*

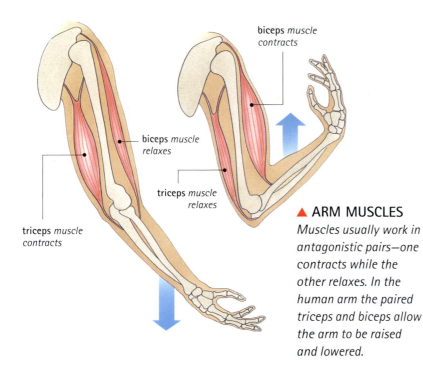

biceps *muscle contracts*

biceps *muscle relaxes*

triceps *muscle relaxes*

triceps *muscle contracts*

▲ **ARM MUSCLES**
Muscles usually work in antagonistic pairs—one contracts while the other relaxes. In the human arm the paired triceps and biceps allow the arm to be raised and lowered.

other. For example, the brachialis and biceps muscles in the upper arm, which make the arm flex, work against the triceps muscle, which extends the arm.

Muscles in the head

There are 10 major muscle groups around the human head and neck. These are involved in supporting and moving the head, making facial expressions, moving the eyes, blinking, speaking, and eating. The facial musculature of chimps and gorillas is similar to that of humans but is not as complex. The strength and location of the muscles that move the lips allow for a uniquely human ability, speech.

A number of muscles in the front part of the neck move the hyoid bone and the voice box, or larynx. These muscles, along with those of the tongue, are used for speech. The principal muscles responsible for chewing are the

CLOSE-UP

Vocal muscles

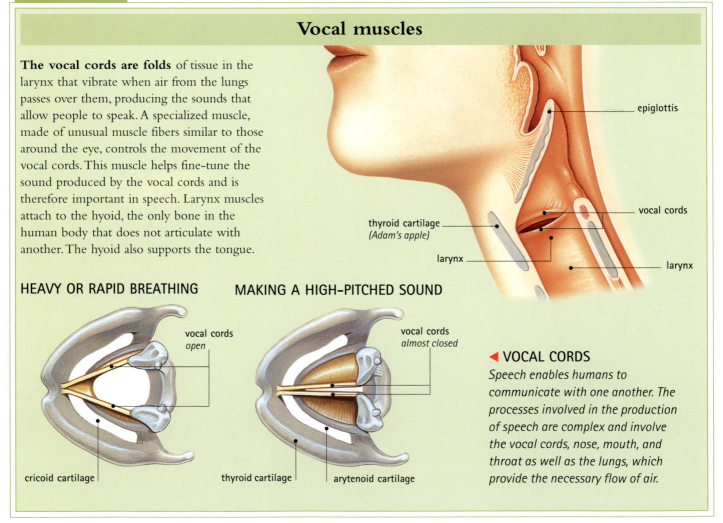

The vocal cords are folds of tissue in the larynx that vibrate when air from the lungs passes over them, producing the sounds that allow people to speak. A specialized muscle, made of unusual muscle fibers similar to those around the eye, controls the movement of the vocal cords. This muscle helps fine-tune the sound produced by the vocal cords and is therefore important in speech. Larynx muscles attach to the hyoid, the only bone in the human body that does not articulate with another. The hyoid also supports the tongue.

epiglottis

vocal cords

thyroid cartilage *(Adam's apple)*

larynx

larynx

HEAVY OR RAPID BREATHING

vocal cords *open*

cricoid cartilage

MAKING A HIGH-PITCHED SOUND

vocal cords *almost closed*

thyroid cartilage

arytenoid cartilage

◄ **VOCAL CORDS**
Speech enables humans to communicate with one another. The processes involved in the production of speech are complex and involve the vocal cords, nose, mouth, and throat as well as the lungs, which provide the necessary flow of air.

The tongue

Relative to its size, the tongue is the strongest muscle in the human body. It needs to be powerful to cope with the large amount of work it must do, including manipulating food and helping produce speech.

masseter muscles. In humans, they are less powerfully developed than those of apes but can still exert considerable pressure.

Muscles of the back and chest

The superficial muscles of the back lie above other deeper muscle layers. There are two main superficial muscle groups: the trapezius in the upper back and the latissimus dorsii in the lower back. These muscles are important for moving the shoulder blades and upper arms, and extending the body from the waist. The arms are also moved by the pectoralis muscles, which extend across the chest.

Below the superficial muscles are the deep back muscles. One group of these muscles, the erector spinae, keeps the spine in an upright position and is important for stabilizing the spine in different positions. There are nine muscles in this group, and they attach to various parts of the dorsal (back) skeleton, including the ribs, the pelvis, and extensions of the vertebrae. The erector spinae muscles are larger in humans than in apes and are critical in maintaining an bipedal stance.

Muscles of the legs

The musculature of human legs is complex; it needs to perform the tricky task of maintaining balance in a bipedal animal with a high center of gravity. Around the buttocks and hips are the gluteal muscles. This group is important for flexing, rotating, and extending the legs. It includes the largest muscle in the human body, the gluteus maximus. Another large member of this muscle group, the well-developed gluteus medius, is crucial for efficient bipedal walking. This muscle supports the side of the pelvis carrying the raised leg during the swing phase of the walking cycle,

when one leg is supporting and balancing the body while the other swings forward. The gluteus medius is poorly developed in chimpanzees; when chimps walk on two legs they must shift the position of their body to maintain balance during the walking cycle, giving them a side-to-side motion.

Toward the back of the thigh is the hamstring muscle group. There are three muscles in this group: the biceps femoris, the semimembranosus, and the semitendinosus. Each of the hamstring muscles is involved in flexing the knee and extending the leg. These muscles are especially active during running. The muscles of the calf in the lower leg, the soleus and gastrocnemius, are particularly large and strong in humans. These muscles give humans the ability to rise on their toes, and provide some of the driving force during walking and running.

Muscle adaptation and training

Human muscles can adapt to the different stresses imposed on them. Athletes carry out different kinds of resistance training to increase muscle mass and improve strength and speed. When muscles are not used they atrophy (lose mass). Astronauts who spend extended periods in low-gravity or zero-gravity conditions lose muscle mass and strength in a matter of days and must exercise regularly to maintain their strength.

Athletes follow rigorous training programs that strengthen the muscles necessary to excel at specific sporting events.

Nervous system

CONNECTIONS

COMPARE the structure of the human eye with that of an invertebrate such as an *OCTOPUS*, whose eyes have evolved independently of vertebrate eyes.

COMPARE the folds on the exterior of the human brain with those of a *MANATEE*. The greater number of folds on the human brain is an indication of the species' greater intelligence.

The nervous system controls and regulates essential and generally unconscious life-support systems, such as heartbeat, temperature regulation, and breathing, as well as coordinating the body's voluntary movements and interpreting the many signals from the sensory organs. The human nervous system is unique in its complexity and is responsible for consciousness, the mind, thought, and people's perception of the world. Like that of other vertebrates, the human nervous system can be divided into the central nervous system (CNS), which includes the brain and the spinal cord, and the peripheral nervous system (PNS), which includes all the nerves and sensory organs attached to the CNS.

The neuron

Neurons are nerve cells. These cells bear long, thin projections called dendrites. Many neurons have one or more long, specialized dendrites called axons. Axons are able to transmit electrical signals rapidly along the length of the cell. These neurons make up some of the longest cells in the body. There are three main types of neurons: sensory neurons, motor neurons, and interneurons. Sensory neurons connect the sense organs to the CNS. Motor neurons carry signals from the CNS to the muscles to effect a response, and interneurons, which occur only in the CNS, connect the other types of neurons.

IN FOCUS

Linking neurons

Junctions between neurons are called synapses. When an electrical signal reaches a synapse, chemicals called neurotransmitters are released into the synapse. The chemicals diffuse across the synapse and join with binding sites on the neighboring neuron. The binding triggers a new signal, allowing the message to pass on.

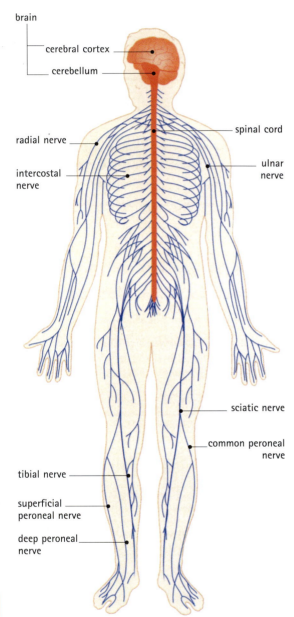

▲ **Human**
A network of nerve fibers—the peripheral nervous system—conveys chemical and electrical messages between all parts of the body and the central nervous system, which is made up of the brain and spinal cord. The central nervous system processes information received from sensory organs and, in response, controls movement, via nerves that connect to muscles, and many other body processes in glands and organs.

Voluntary and involuntary

The PNS can be subdivided into two distinct systems: the somatic nervous system and the autonomic nervous system. The somatic system controls voluntary actions of the skeletal muscles; actions such as walking and talking depend on the somatic nervous system. The autonomic system controls involuntary body processes over which little or no conscious control can be exerted, such as the heartbeat. The autonomic nervous system also triggers the release of secretions from certain glands, such as production of epinephrine by the adrenal glands.

Autonomic nerves continuously regulate internal body conditions. There are two distinct subsystems within the autonomic nervous system: the sympathetic and parasympathetic systems. The sympathetic system governs responses associated with fear, escape, and aggression, such as the release of epinephrine from the adrenal glands and increases in heart rate. The parasympathetic system has the opposite effect; it stimulates tissues associated with digestion and relaxation.

Peripheral organization

The cranial nerves, which include the optic nerves, exit the brain directly and form a major section of the PNS. There are also 31 pairs of spinal nerves that branch off from the spinal cord. Nerves are bundles of sensory and motor axons. The cell bodies of motor neurons, which contain the nucleus and other cellular machinery, lie in the spinal cord. Their axons pass out of the spinal cord on the front (ventral) side of the spinal column. Sensory neurons pass out of the back of the spinal cord. The cell bodies of sensory neurons occur in a cluster (or ganglion) outside the spinal

CLOSE-UP

Human vision

The most complex sensory organs in humans are the eyes. Humans rely heavily on vision, and are among relatively few mammals that can see colors. The eyes account for around 1 percent of the total head weight. Light enters the eye through the cornea, passes through the iris, and then through a lens. The lens focuses an inverted image onto a network of light-sensitive cells at the back of the eye called the retina. They then fire electrical signals to the brain through the optic nerves; the signals are converted into an image. In humans, the visual lobe of the brain is much larger than the olfactory lobe, which processes the sense of smell, reflecting humans' greater reliance on vision.

▶ EYEBALL

Cells at the back of the eye called cones and rods detect color and monochrome light respectively. Nerve impulses from these cells are interpreted by the brain as images.

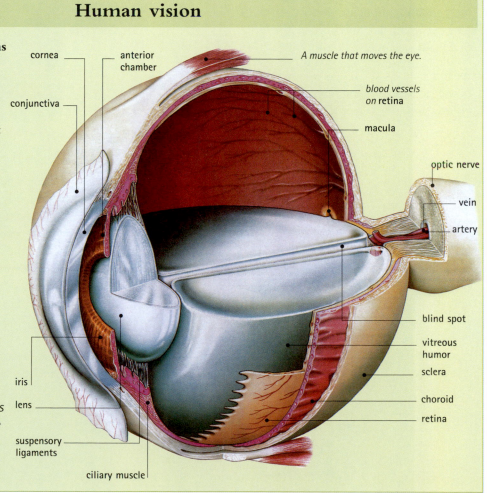

cornea
anterior chamber
A muscle that moves the eye.
conjunctiva
blood vessels on retina
macula
optic nerve
vein
artery
blind spot
vitreous humor
sclera
choroid
retina
iris
lens
suspensory ligaments
ciliary muscle

Brain size and smartness

Various methods have been used to attempt to measure the relative intelligence of animals. One method is the encephalization quotient (EQ). This is a number arrived at using a mathematical formula that involves comparing brain mass with body mass. A mammal with an EQ value of 1 represents a brain of expected size, and therefore average intelligence. Humans have by far the highest EQ (around 7.44). Neanderthals had an even larger brain relative to their size. Dolphins are second in the "smart list" of living animals (5.31), far ahead of chimpanzees (2.49). Mice, rats, and rabbits have the lowest EQ, with brains less than half the size expected for animals of their size.

Another method for assessing relative intelligence involves measuring the degree of folding in the cerebral cortex, the most recently evolved area of the brain and the one most often linked with the development of intelligence.

Such methods as these, however, are controversial and are not thought to provide entirely accurate results.

The brain
The human brain is divided into three main parts: the brain stem, the cerebellum, and the cerebrum.

The frontal lobe is involved in complex thinking.

Region of the cortex that processes speech.

Areas controlling body movements.

The cerebrum; the folded outer layer is called the cortex.

touch

parietal lobe

taste

vision

occipital lobe

cerebellum

brain stem

temporal lobe

Region of the cortex responsible for hearing.

Region of the cortex that processes smell.

---

cord. The axons from the sensory and motor neurons join into a single nerve; they branch again farther along the nerve.

The central nervous system
The spinal cord of adult humans measures around 18 inches (45 cm) long, and is about as thick as a person's thumb. The brain weighs around 3 pounds (1.4 kg) and is made up of at least 10 billion neurons. The entire CNS is surrounded by a liquid called the cerebrospinal fluid (CSF), which protects the delicate tissues and transports of oxygen and nutrients.

A single neuron in the brain can connect with 100,000 or more other neurons, with different rules governing the transmission of electrical signals between different neurons. The complexity of the human brain is crucial for the production of a unique level of consciousness and the capacity for abstract

thought. Different parts of the brain are associated with different tasks, such as speech and body coordination. However, like all other vertebrate brains the human brain consists of three main parts: the forebrain, the midbrain, and the hindbrain. The hindbrain contains the medulla and pons, and the cerebellum and olfactory bulb are part of the midbrain. These regions are associated with life-support functions. The medulla regulates breathing and heart rates, blood pressure, and digestion. The cerebellum helps control balance, posture, and muscle coordination. These parts of the brain are the first to develop in fetal mammals and are collectively called the archipallium.

The paleopallium
The next part of the brain to form during the growth of the fetus is the paleopallium, which mostly consists of the limbic system. This

includes the amygdala, which is responsible for emotions such as fear and anger; and the hippocampus, which is involved in memory and learning. Other parts of the limbic system also control emotions, moods, and motivations such as sexual drive.

The thalamus and hypothalamus are important paleopallium structures. The thalamus is involved in the coordination of movement and the processing of sensory information. It passes information to and from the most developed part of the human brain, the neocortex. The hypothalamus controls body temperature, the sensations of hunger and thirst, and the internal body clock; it also controls hormone output from the pituitary gland.

The neocortex

The neocortex makes up much of the cerebral hemispheres. It is a relatively recently evolved structure and only occurs in mammal groups such as ungulates, cetaceans, carnivores, and

primates. The folding of the cortex creates a series of bumps and grooves called gyri and sulci. More gyri and sulci increase the brain's surface area. Particular regions of the neocortex are associated with advanced brain functions such as language, reasoning, perception, and voluntary movement. The neocortex is the seat of human consciousness.

IN FOCUS

Parkinson's disease

People suffering from Parkinson's disease have difficulty moving; they lack coordination and experience serious tremors. This condition is not caused by any problems of the muscles but is due to the loss of neurons in the substantia negra, part of the brain important for coordinating movement. These neurons produce a neurotransmitter called dopamine. As levels of dopamine fall, amounts of another brain chemical, acetylcholine, increase. It is this chemical imbalance that causes the tremors that characterize Parkinson's disease.

STRUCTURE OF THE EAR

HEARING MECHANISM

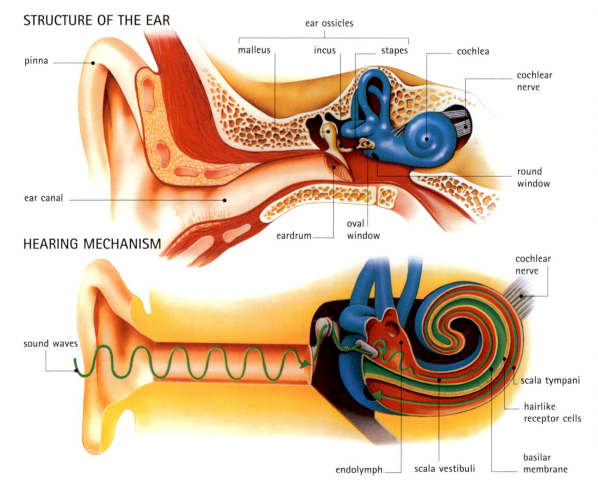

ear ossicles: malleus, incus, stapes, cochlea
pinna
cochlear nerve
ear canal
round window
eardrum
oval window
sound waves
cochlear nerve
scala tympani
hairlike receptor cells
basilar membrane
endolymph
scala vestibuli

◀ EAR

Sound waves travel along the ear canal causing the eardrum to vibrate. These vibrations are transferred to the ear ossicles, which in turn cause a membranous structure called the oval window to vibrate. The sound pulses are transferred by the oval window to the endolymph in the scala vestibuli. They then pass across the basilar membrane, where hair-like receptor cells detect the pulses and produce nervous impulses, which are interpreted by the brain as sounds. The vibrations then pass to the scala tympani and on to the round window.

Circulatory and respiratory systems

Oxygen is required for cellular respiration, the series of chemical reactions that converts nutrients from food into usable energy. Small organisms such as amoebas and tiny worms can get enough oxygen by allowing it to diffuse through their outer membranes into the cell body. However, this diffusion becomes increasingly ineffective as animals get larger. Larger animals need dedicated organs to ensure that enough oxygen reaches all the cells in their body.

Respiratory system

Like almost all other tetrapods (four-limbed vertebrates) humans use a pair of lungs for bringing oxygen into the bloodstream and removing carbon dioxide, the waste product of cellular respiration. Air enters the lungs through the nose or the mouth; passes down the windpipe, or trachea; and then passes into a pair of bronchi. The bronchi split into thousands of smaller tubes called bronchioles.

EVOLUTION

Life at high altitudes

At altitudes above about 10,000 feet (3,000 m) the air is thin and has little oxygen. Visitors to these high altitudes grow dizzy and feel sick owing to lack of oxygen. Their breathing rate rises, and if they stay longer than a few weeks they begin to make more hemoglobin, the protein that carries oxygen in the blood. However, people who live in mountain ranges such as the Andes and the Himalayas have adapted for life at high altitudes in apparently different ways. Scientists have discovered that people living in the Andes have higher concentrations of oxygen-carrying hemoglobin in their blood but breathe at the same rate as people living at low altitudes. People in the Himalayas of Tibet, however, breathe more quickly to obtain more oxygen. Both adaptations for life at high altitudes are successful.

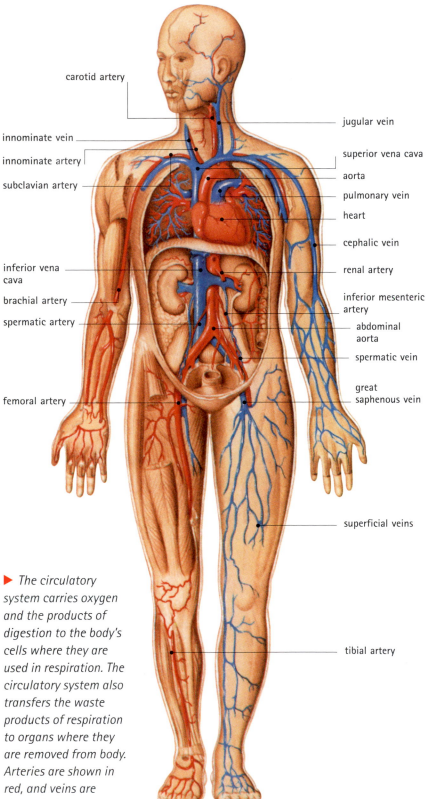

carotid artery

innominate vein

innominate artery

subclavian artery

inferior vena cava

brachial artery

spermatic artery

femoral artery

jugular vein

superior vena cava

aorta

pulmonary vein

heart

cephalic vein

renal artery

inferior mesenteric artery

abdominal aorta

spermatic vein

great saphenous vein

superficial veins

tibial artery

▶ The circulatory system carries oxygen and the products of digestion to the body's cells where they are used in respiration. The circulatory system also transfers the waste products of respiration to organs where they are removed from body. Arteries are shown in red, and veins are shown in blue.

All these respiratory tubes are reinforced with rings of cartilage to keep them from collapsing. At the end of the bronchioles are small air sacs called alveoli, where gas exchange occurs. There are millions of alveoli in human lungs; they give a total surface for respiration of more than 800 square feet (75 m²).

Unlike other tetrapods, humans and other mammals breathe using a large muscle called the diaphragm, which forms a muscular floor below the ribs. The diaphragm increases the volume of the chest cavity when it contracts, thus decreasing the pressure in this cavity and causing air from outside the body to inflate the lungs. Humans are one of the few terrestrial mammals that can voluntarily hold their breath. This ability is assisted by the unusually low position of the larynx in the trachea and probably evolved as an aid to speech.

The circulatory system
Mammals and birds can control their body temperature internally, but this requires much more energy than a cold-blooded physiology, such as that of reptiles. Mammals and birds therefore need an efficient circulatory system to supply their oxygen-hungry tissues.

IN FOCUS

How heart attacks happen

Heart attacks can occur when the arteries supplying oxygenated blood to the muscles of the heart become blocked or inflexible. When this blocking or hardening happens, the oxygen supply to parts of the heart is cut down and the deprived sections can eventually fail, leading to the pain and problems of a heart attack. The changes to the artery walls that cause this are called arteriosclerosis, or hardening of the arteries. The damaging changes are partly a loss of elasticity in the artery wall and partly a laying down of fatty deposits inside the artery, blocking blood flow much as lime scale narrows a water pipe. These problems tend to accumulate with age, but can be made worse by smoking and a diet high in animal fats. High levels of cholesterol in the blood can lead to this kind of arterial blockage.

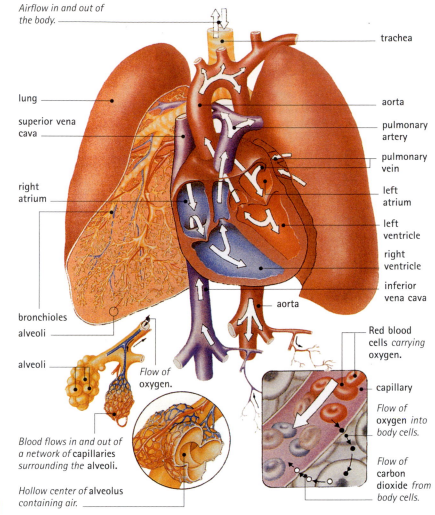

Airflow in and out of the body.

trachea

lung

aorta

superior vena cava

pulmonary artery

pulmonary vein

right atrium

left atrium

left ventricle

right ventricle

inferior vena cava

aorta

bronchioles

alveoli

Red blood cells *carrying* oxygen.

alveoli

Flow of oxygen.

capillary

Flow of oxygen into body cells.

Blood flows in and out of a network of capillaries surrounding the alveoli.

Flow of carbon dioxide from body cells.

Hollow center of alveolus containing air.

Birds and mammals pump blood around their body using a four-chamber heart, consisting of two atria and two ventricles. Blood loaded with carbon dioxide and depleted of oxygen first enters the right atrium, which squeezes blood through a set of valves into the right ventricle. Contraction of the right ventricle then pushes blood through one-way valves into the pulmonary artery, which leads to the lungs. The pulmonary artery branches into thousands of tiny capillaries that wrap around the alveoli; exchange of carbon dioxide and oxygen takes place across the surfaces of the capillaries and alveoli. Oxygen-rich blood returns to the left atrium of the heart via the pulmonary veins. Blood is squeezed through valves into the most powerfully built chamber of the heart, the left ventricle. Contractions from this chamber force blood into the body's main artery, the aorta, from which other arteries stem.

▲ HEART AND LUNGS
Air passes along the trachea and into the bronchi of the lungs. It then passes into smaller tubes called bronchioles until it reaches tiny air sacs called alveoli. There oxygen in the air diffuses across the walls of the alveoli into tiny blood vessels called capillaries. Red blood cells pick up the oxygen and carry it to the body's cells. Carbon dioxide takes the reverse path out of the body.

Digestive and excretory systems

COMPARE the human cecum and appendix with the cecum of a herbivore such as a **HARE**.

COMPARE the intestines of an omnivorous human with the intestines of a herbivore such as a **GIRAFFE** and a carnivore such as a **LION**.

CONNECTIONS

Vertebrate digestive systems generally can be divided into four main parts. These are the buccal cavity (inside the mouth) and the associated food pipe (the esophagus); the stomach; the small intestine; and the large intestine, which leads to the anus. As in other vertebrates, humans possess a liver that secretes digestive chemicals and processes the nutrients absorbed through the intestine wall, and a pair of kidneys to extract and excrete superfluous or toxic substances from the blood.

Food's journey
Food is swallowed and forced down the esophagus by muscular contractions until it reaches the stomach. The stomach is an elastic, muscular bag that can stretch greatly after a large meal. In the stomach lining, or epithelium, there are many gastric pits lined with cells that secrete hydrochloric acid and protein-digesting enzymes. Goblet cells in the lining of the stomach produce copious mucus secretions, which protect the stomach wall from the strong acid.

Compared with the large multichambered stomachs of ruminants like cattle, the human stomach is simple, like that of most carnivores. Semi-digested food leaves the stomach in small portions and moves into the duodenum, the first 10 inches (25 cm) of the small intestine. There, further protein and carbohydrate digestion takes place, a process driven by a cocktail of enzymes secreted from glands in the epithelium of the duodenum and the pancreas. Bile salts produced in the liver and stored in the gallbladder are added to the mixture to break up fats into small droplets to aid fat digestion. Movement of food is driven by waves of muscular action called peristalsis.

Maximum area
The rest of the small intestine consists of sections called the jejunum and the ileum. These regions are devoted to food absorption,

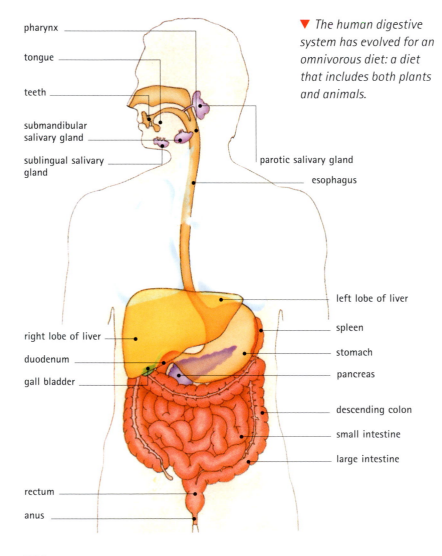

▼ The human digestive system has evolved for an omnivorous diet: a diet that includes both plants and animals.

pharynx, tongue, teeth, submandibular salivary gland, sublingual salivary gland, parotic salivary gland, esophagus, left lobe of liver, spleen, right lobe of liver, stomach, duodenum, pancreas, gall bladder, descending colon, small intestine, large intestine, rectum, anus

IN FOCUS

The liver

After the skin, the liver is the largest organ in the human body, weighing around 3 pounds (1.4 kg). The liver is responsible for making proteins, storing energy in the form of the starch glycogen, storing vitamins and minerals, and breaking down toxins. Toxin breakdown is particularly important in humans because the rapid passage of food through the stomach followed by prolonged retention in the colon may make humans susceptible to bacterial infections. Alcohol is broken down in the liver by the enzyme alcohol dehydrogenase.

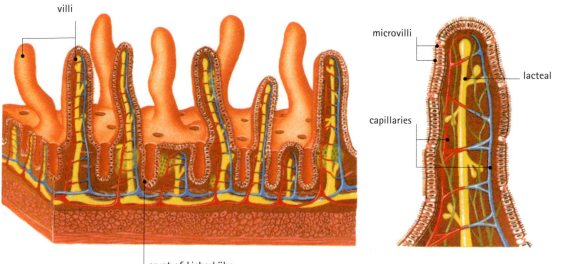

villi

crypt of Lieberkühn

microvilli

lacteal

capillaries

In the duodenum, small projections called villi absorb the products of digestion. These projections dramatically increase the surface area of the intestine, allowing the rapid and efficient absorption of digested food. Small depressions between the villi, called crypts of Lieberkühn, secrete digestive juices.

with no further enzyme secretion. Most meat eaters have relatively short small intestines, ranging from two to six times the body length. The human small intestine falls within this range. Herbivore small intestines are much longer; those of ungulates may be between 20 and 25 times longer than the body.

The surface area of the inside of the small intestine is very large; this area is important for maximizing the rate of absorption. The surface area is increased through folds called villi. These are covered by tiny fingerlike projections called microvilli. The average surface area of the small intestine of an adult human is around 3,350 square feet (310 m²).

COMPARATIVE ANATOMY

Teeth

Adult humans have 32 teeth, which include eight incisors, four canines, eight premolars, and 12 molars. The incisors have straight edges that are useful for nipping, and the molars and premolars are used for mashing food. Human teeth are not adapted for slicing meat like the carnassial teeth of carnivores, nor are they adapted for grinding a heavy fibrous vegetable diet, like the ridged molars of many ungulates. Gorillas are vegetarian and have the most powerfully built molars of the great apes. Gorillas and chimpanzees also have large canines, but these are adaptations for fighting and display rather than feeding.

Into the large intestine

The small intestine opens into the cecum at the beginning of the large intestine. At only 5 to 6 feet (1.5–1.8 m) long, the large intestine is much shorter than the small intestine, but it is much thicker and bulkier. Some herbivorous animals such as horses and rabbits have an enlarged cecum where the bacterial fermentation of vegetation takes place. Humans have a small cecum, with an outgrowth called the appendix. This outgrowth serves no function in modern humans; it is a vestigial structure. Some bacterial fermentation does take place in the large intestine, especially after certain meals such as beans, resulting in gaseous emissions. In humans, the large intestine is primarily the site of water and vitamin absorption and a storage place for feces, the remaining undigested material, prior to release at the anus.

◄ Digestion begins in the mouth, where salivary amylase begins the process of breaking down starches to simple sugars.

Endocrine and exocrine systems

CONNECTIONS

COMPARE the mammary glands of a human with those of a monotreme mammal such as a *PLATYPUS* and a marsupial such as a *KANGAROO*.

Glands are tissues that secrete a variety of important substances, ranging from sweat to hormones. The glands of the exocrine system are connected to ducts that channel the secreted substance to a surface. These glands include the sweat and mammary glands, which secrete sweat and milk, respectively, to the surface of the skin; and the pancreas, liver, and glands in the wall of the gut, which secrete digestive enzymes and bile onto the inner surface of the digestive tract. Endocrine glands have no ducts; they secrete their products into the bloodstream. Endocrine glands produce chemical messengers, or hormones. Examples of endocrine glands include the ovaries, testes, adrenal glands, and pituitary gland.

▶ **FEMALE ENDOCRINE SYSTEM**

The endocrine system consists of glands that secrete chemical messengers called hormones into the bloodstream. In contrast, exocrine glands secrete substances to a surface such as the surfaces of the stomach lining and the skin.

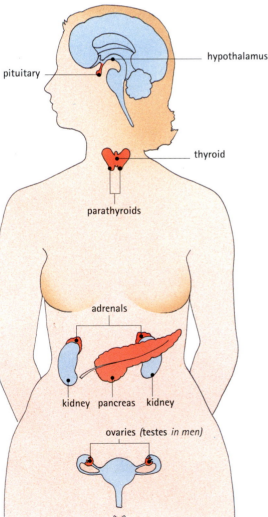

hypothalamus

pituitary

thyroid

parathyroids

adrenals

kidney   pancreas   kidney

ovaries *(testes in men)*

The exocrine system

Some exocrine glands such as the salivary glands, liver, and pancreas occur in all vertebrates, although they may take different forms. Humans have three pairs of salivary glands that secrete enzymes in a solution of mucus; the salivary glands lubricate food and begin the digestive process. Vertebrate liver tissue is organized into small lobes, consisting of strings of liver cells surrounding a central vein coming from the small intestine. Bile ducts surround each lobe to channel away bile and waste products excreted by the cells. All four-limbed vertebrates have a two-lobed pancreas.

Uniquely mammalian

Some exocrine glands, such as the sweat and mammary glands, occur only in mammals. The sweat glands secrete liquids onto the surface of the skin. They help cool the animal; the sweat draws away heat as it evaporates. Not all mammals have sweat glands. Sweat glands are absent in some marine mammals, and in mammals such as dogs they are restricted to just a few places. That is why a hot dog pants; it loses body heat through its tongue.

IN FOCUS

Fight or flight

The adrenal glands, situated above each kidney, are vital for the "fight or flight" response to danger in a human or another vertebrate. These orange glands contain a central area, the medulla, and a surrounding layer, called the cortex. When signals from the sympathetic nervous system reach the adrenal medulla two hormones are produced. These are epinephrine (adrenaline) and norepinephrine. These hormones stimulate a number of responses that assist immediate action, including increased heart rate, dilation of airways in the lungs to increase oxygen uptake, and restricted gut action so more energy can be focused on the muscles.

Sugar levels

The level of glucose, a sugar, in the blood is kept under strict control and must be maintained within a very narrow range. Glucose is controlled by secretions from part of the pancreas called the islets of Langerhans, which produces the hormones insulin and glucagon. If there is too much sugar in the blood, the pancreas secretes insulin, which helps cells absorb the excess. If there is too little sugar, glucagon is released; this triggers the liver to release glucose from storage.

Some people with diabetes control the disease with injections of insulin to regulate their blood sugar levels.

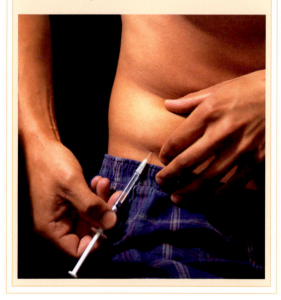

Human females have two mammary glands that produce nourishing milk for newborn offspring. It is rare for a woman to give birth to more than two babies at once, so two mammary glands are generally sufficient. Mammals that produce larger numbers of young have many more mammary glands.

The endocrine system
One of the most important endocrine glands is the pea-size pituitary gland located just under the hypothalamus of the brain. The pituitary controls the secretion of hormones by other endocrine glands. It has two distinct parts: the posterior pituitary, which derives

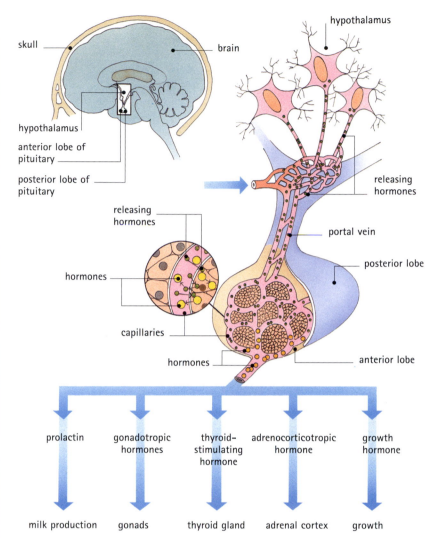

from the hypothalamus; and the anterior pituitary, which is formed by tissues originating from the roof of the mouth. The posterior pituitary is connected to the hypothalamus by neurons and secretes two hormones: antidiuretic hormone (ADH), which regulates water balance; and oxytocin, which promotes muscular contractions in the mammary glands during lactation and in the uterus during childbirth.

The anterior pituitary produces more hormones. Most stimulate hormone secretion in other endocrine glands. For example, the anterior pituitary produces luteinizing hormone; this stimulates the testes to produce testosterone. Testosterone is a hormone that starts and maintains male characteristics such as body hair and large muscle size. In females, luteinizing hormone plays an important role in the menstrual cycle.

▲ **HYPOTHALAMUS**
The hypothalamus secretes hormones that pass to the pituitary gland and cause the pituitary to release further hormones, which are secreted into the bloodstream. Most of the hormones released by the pituitary gland stimulate other glands to release still further hormones.

Reproductive system

CONNECTIONS

COMPARE human males' reproductive organs with those of a *HARE*, in which the position of the penis and testicles is reversed.

COMPARE the position of the human female's vagina with that of a female *ELEPHANT*. The elephant's vagina is positioned on the underside of the animal, nearer the stomach.

Men produce sperm in a pair of testes; a woman's eggs are produced in a pair of ovaries. When compared with many other mammals, humans have some unusual sexual strategies and structures that are a legacy of our species' evolutionary history.

The male reproductive system

The body temperature of mammals is too warm for optimum sperm production, so the testes of men and most other mammals are held outside the body cavity in a sac called the scrotum. The temperature of the scrotum is usually around 7.2°F (4°C), cooler than inside the body. Muscles in the scrotum move the testes slightly in response to changing temperature. Inside each testis, sperm is stored in a coiled tube called the epididymis.

Male mammals and many other animals insert sperm into the female with an organ called the penis. Mammal penises develop around the urethra, the tube that transports urine from the bladder. The penis contains chambers filled with spongy material that become engorged with blood during sexual arousal. The blood-filled chambers provide the rigidity that allows copulation to take place.

The female reproductive system

The ovaries are inside the body cavity of the female. Mammal oviducts (tubes that lead to the outside) form several discrete structures. The fallopian tubes lead from the ovaries to the uterus, where the fertilized egg develops into an embryo. The uterus connects through the cervix to the vagina, an elastic tube which receives the penis during copulation and through which young are born.

During ovulation, an egg is released from the ovary to the fallopian tube. The walls of the uterus thicken at this time. If the egg remains unfertilized, the uterus walls are shed in a process called menstruation.

Copulation and fertilization

During copulation, the male releases up to 400 million sperm into the vagina. The sperm pass from the epididymis into tubes called the vasa

▶ **MALE REPRODUCTIVE SYSTEM**

The male reproductive organs are able to produce sperm and deposit them in the female reproductive organs. Erectile tissue in the penis enables the penis to become stiff so that it may be inserted in the vagina. Sperm is made in the testes and stored in the epididymis.

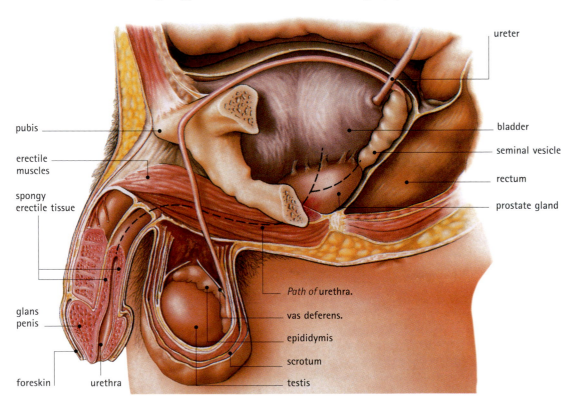

pubis
erectile muscles
spongy erectile tissue
glans penis
foreskin
urethra

ureter
bladder
seminal vesicle
rectum
prostate gland
Path of urethra.
vas deferens.
epididymis
scrotum
testis

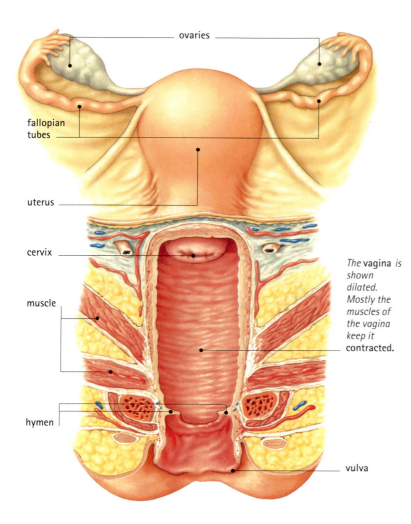

ovaries

fallopian tubes

uterus

cervix

muscle

hymen

The vagina is shown dilated. Mostly the muscles of the vagina keep it contracted.

vulva

▲ FEMALE
REPRODUCTIVE
SYSTEM

The race for fertilization

Male mammals compete for females before copulation, and sperm also compete inside the female in the race for fertilization. This sperm competition can have a major effect on male reproductive anatomy. For example, the size of primate testes varies depending on the type of society that the animals live in. There is a neat correlation between testis size and levels of promiscuity, the number of sexual partners an animal has. Having many different partners implies more rivals' sperm for an individual male to compete with. Bigger testes produce more sperm with which to overwhelm the sperm of rivals. This is a vital advantage.

Promiscuity is almost unknown in gorillas, and they have correspondingly tiny testes. However, bonobo chimpanzees are extremely promiscuous. Female bonobos mate up to 1,000 times per fertilization; male bonobos have very large testes. Male humans' testes are an intermediate size.

These lines of evidence suggest that recent human ancestors lived in societies with moderate amounts of promiscuity. Humans remain, to an extent, promiscuous. Between 2 and 30 percent of all babies are the products of extra-pair paternity; they are fathered by a man from outside the parental couple.

Development of the embryo

A fertilized egg moves from the fallopian tube to the uterus, carried by muscular contractions of the tube. Filaments called cilia on the inside of the tube also help transport the egg. The egg embeds into the thickened uterus wall, where it develops into an embryo. The embryo is nourished and provided with oxygen through an temporary organ called the placenta. This grows partly from the embryo and partly from the mother. Pregnancy ends after around nine months, when the baby is born through the vagina.

ADRIAN SEYMOUR

deferentia, which lead to the urethra. Glands, such as the prostate gland, secrete other components of the seminal fluid, or semen. These secretions carry the sperm and contain a sugar called fructose that nourishes them.

EVOLUTION

Sexual dimorphism

Like many mammals, male and female humans are different shapes and sizes. This difference is called sexual dimorphism. Generally, men are larger and heavier; this may be an adaptation for fighting rivals. Females have wider hips that result from a differently shaped pelvis. Wider hips accommodate childbirth. Among the more striking sexually dimorphic traits are women's breasts. Female chimps and other primates suckle their young perfectly well without having breasts; if anything, breasts hinder the action of suckling by human babies. This suggests that human breasts evolved for display to males, perhaps as an indicator of health or to stimulate sexual activity.

FURTHER READING AND RESEARCH
Baggaley, A. and J. Hamilton. 2001. *Human Body: An Illustrated Guide to Every Part of the Human Body and How It Works.* DK Publishing: NY.
Van der Graaf, K. 1997. *Schaum's Outline of Human Anatomy and Physiology.* McGraw-Hill: Columbus, OH.

Index